GROWTH FACTORS IN REPRODUCTION

SERONO SYMPOSIA, USA

FERTILIZATION IN MAMMALS
Edited by Barry D. Bavister, Jim Cummins, and Eduardo R. S. Roldan

GAMETE PHYSIOLOGY
Edited by Ricardo H. Asch, Jose P. Balmaceda, and Ian Johnston

GLYCOPROTEIN HORMONES: Structure, Synthesis, and Biologic Function
Edited by William W. Chin and Irving Boime

GROWTH FACTORS IN REPRODUCTION
Edited by David W. Schomberg

THE MENOPAUSE: Biological and Clinical Consequences of Ovarian Failure: Evaluation and Management
Edited by Stanley G. Korenman

NEUROENDOCRINE REGULATION OF REPRODUCTION
Edited by Samuel S. C. Yen and Wylie W. Vale

UTERINE CONTRACTILITY: Mechanisms of Control
Edited by Robert E. Garfield

Serono Symposia, USA
Norwell, Massachusetts

GROWTH FACTORS IN REPRODUCTION

Edited by

DAVID W. SCHOMBERG

Duke University Medical Center
Durham, North Carolina

Springer-Verlag
New York Berlin Heidelberg London
Paris Tokyo Hong Kong Barcelona

David W. Schomberg, PhD
Duke University Medical Center
Durham, NC 27710
USA

Proceedings of the Symposium on Growth Factors in Reproduction, sponsored by Serono Symposia, USA, held April 1 to 4, 1990, in Savannah, Georgia.

For information on previous volumes, please contact Serono Symposia, USA.

Printed on acid-free paper.

CR copy provided by Technical Texts, Inc., Scituate, Massachusetts.

9 8 7 6 5 4 3 2 1

ISBN-13: 978-1-4612-7819-1 e-ISBN-13: 978-1-4612-3162-2
DOI: 10.1007/ 978-1-4612-3162-2

SYMPOSIUM ON GROWTH FACTORS IN REPRODUCTION
Scientific Committee

David W. Schomberg, Ph.D., *Chairman*
Durham, North Carolina

Eli Y. Adashi, M.D.
Baltimore, Maryland

Robert B. Dickson, Ph.D.
Washington, D.C.

David C. Lee, Ph.D.
Chapel Hill, North Carolina

John A. McLachlan, Ph.D.
Research Triangle Park, North Carolina

R. Michael Roberts, Ph.D.
Columbia, Missouri

Organizing Secretary

L. Lisa Kern, Ph.D.
Serono Symposia, USA
100 Longwater Circle
Norwell, Massachusetts

Preface

The critical role of growth factors in normal as well as abnormal cellular growth and function has become overwhelmingly apparent in recent years. The significant advances in reproductive biology within this rapidly expanding field were highlighted at the Serono Symposia, USA, symposium entitled Growth Factors in Reproduction. The conference focused on growth factors as polyfunctional regulators of growth and development in the reproductive system.

The program was organized into five areas: (1) growth factors as polyfunctional regulators of growth and development, (2) growth factors and gonadal function, (3) regulation of normal and neoplastic mammary growth, (4) regulatory peptides in reproductive tract development and function, and (5) embryo-maternal signaling. This volume constitutes the contributions of the invited symposium speakers and is organized into sections representing the five topic areas of the sessions.

Drs. Eli Y. Adashi, Robert B. Dickson, David C. Lee, James Hammond, John A. McLachlan, and R. Michael Roberts, who served as chairmen of these sessions, not only provided invaluable advice and assistance in organizing the symposium, but contributed significantly to its content. The enthusiasm and esprit de corps of the poster presenters and discussion participants also helped make this symposium succeed as the first of its kind.

Finally, special thanks are due Drs. James Posillico and Lisa Kern of Serono Symposia, USA, for their guidance and support in the conception and planning of the symposium and in the publication of this volume.

David W. Schomberg

Contents

PART

I

GROWTH FACTORS: POLYFUNCTIONAL REGULATORS OF GROWTH AND DEVELOPMENT

1

The Epidermal Growth Factor Receptor: Control of Synthesis and Signaling Function

Shelton Earp, William Huckle, Victoria Raymond, Leslie Petch, Sherry Marts, Warren Bishop, and Bryan McCune

Lineberger Cancer Research Center, University of North Carolina at Chapel Hill

This chapter serves two functions. The first is to introduce the area of growth factor receptors and signal transduction to an audience well versed in the action of reproductive tract hormones. The second is to summarize work from our own lab regarding the regulation of synthesis and function of the rat epidermal growth factor (EGF) receptor.

Researchers from a variety of disciplines have used molecular and cellular techniques to study the molecules involved in the regulation of cell growth. Growth has been investigated in virtually every organ system and, as a number of specific growth factors and growth factor receptors have been discovered, general principles of growth factor and receptor action are emerging (1–2). In part, these general precepts are derived from structural analyses made possible by sequencing growth factor receptors. In addition, biochemical studies have demonstrated the general features of postreceptor signal transduction. To orient the reader, an oversimplified schema is presented below. Proliferation is initiated by growth factor binding to an inactive receptor. This results in transmission of a physical signal via the receptor through the cell's plasma membrane. Most often, this signal results in activation of a tyrosine-specific protein kinase. This kinase may be encoded within the cytoplasmic domain of the receptor itself (1–2), or it may be an intracellular tyrosine kinase linked to a receptor by noncovalent interactions. (For example, T-cell antigen receptor-stimulated growth of T-cells may be mediated in part by an intracellular tyrosine kinase, *fyn* [3–4].) In either case, there is an immediate burst of intracellular tyrosine phosphorylation following ligand binding. This in turn influences other signal transduction systems (i.e., second messengers) as well as the cell's cytoskeletal structure. This network of signals eventually affects the expression of specific genes, setting in motion a program that results in cell growth. While some growth-promoting agents do not act via tyrosine phosphorylation, the tyrosine kinase-linked growth factor receptors predominate and are the focus of this review. In the second part of this chapter, we detail studies of the initial burst of tyrosine phosphorylation produced by EGF in liver cells and the consequences of this signaling on the regulation of EGF receptor gene expression. While not the focus of this chapter, it should be recognized that hormones can (*a*) prepare a tissue for growth, (*b*) play an important permissive role in growth factor action, or (*c*) induce the production or secretion of growth factors.

OVERVIEW OF GROWTH FACTOR RECEPTOR ACTION

The study of growth factor and growth factor receptor action precedes the current interest in cellular oncogenes. Dr. Stanley Cohen began his pioneering analyses of EGF, one of the first well-characterized growth factors, in the 1950s and 1960s (5–6). In the 1970s the studies of tumor viruses demonstrated that oncogenes encoded by retroviruses were mutated forms of normal cellular genes termed *proto-oncogenes* (7). This was followed by the realization that several proto-oncogenes were, in fact, growth factors or growth factor receptors (8). The hypothesis emerged that a cascade beginning with growth factors and their receptors regulates intracellular signal transduction and, in turn, controls gene expression. The consequence of mutation in one or more of the constituents of the growth control cascade may result in a persistent growth signal that initiates neoplastic transformation.

The synthesis of the fields of growth factor and oncogene research came in the early 1980s with the identification of the simian sarcoma virus *sis* oncogene as the B-chain of *platelet-derived growth factor* (PDGF), a mitogen released from platelets after aggregation (9–11). Subsequently, the isolation and sequencing of the human EGF receptor demonstrated that it was the proto-oncogene from which the v-*erb* B oncogene arose (12–13). Thus, v-*erb* B was a truncated form of the EGF receptor that lacked the external ligand-binding domain. The cell's exposure to a constitutively activated EGF receptor resulted in a malignant phenotype. More recently, other oncogenes isolated from viruses or by transfection have subsequently proved to be mutated forms of transmembrane proteins exhibiting tyrosine kinase activity intrinsic to their cytoplasmic domains; for example, *fms*, *neu*, *met*, *kit*, *ros*, and *trk* (2).

Tyrosine-Specific Protein Kinases

Serine and threonine protein kinases were known to be important members of second-messenger pathways. Analyses demonstrated that tissues have greater than 99% of their protein-bound phosphate on serine and threonine residues (14). While tyrosine phosphate was a known chemical entity, its biologic relevance was undefined until the discovery that the v-*src* (15) and v-*abl* (16) oncogenes exhibited tyrosine-specific protein kinase activity. The initial studies showed that v-*src* immunoprecipitates contained an intrinsic protein kinase activity (17). Surprisingly, phosphoamino acid analysis showed that the kinase activity resulted in phosphorylation of tyrosine (15). Thereafter, it has been shown that a number of retroviral oncogenes exhibit tyrosine kinase activity (18). The interest in this kinase family was heightened by the discovery that the EGF receptor also exhibited intrinsic tyrosine kinase activity, an activity that was stimulated by the addition of EGF (19–20).

The *src* oncogene and the EGF receptor still typify the two major types of tyrosine kinases. The first type (EGF receptor-like) is intrinsic to transmembrane receptors and is presumably activated by extracellular perturbation. Tyrosine kinases of the second type (*src*-like) are intracellular molecules that are posttranslationally modified (e.g., by the addition of myristic acid) such that they bind to the inner leaflet of membranes, particularly the plasma membrane (21). There is probably a third class of tyrosine kinases that is cytosolic. Tyrosine kinases are identifiable by a region of amino acid similarity spanning ~290 residues (18). This region also exhibits sequence similarity to serine/threonine protein kinase families, but certain invariant amino acids clearly distinguish the tyrosine from serine/threonine kinases (18, 22). More than 35 tyrosine kinases in these two classes have been cloned and sequenced (22). With the advent of PCR-based cloning techniques, the addition of new kinase sequences is likely to accelerate. However, in spite of this plethora of tyrosine kinases, cells maintain their phosphotyrosine content at less than 1% of total protein-bound phosphate. The importance of tyrosine kinases to the control of growth and development is emphasized by the fact that most sequenced growth factor receptors are members of the tyrosine kinase family (2). In addition, many proteins involved in the control of development in lower organisms (e.g., *Drosophila*) are also tyrosine kinases (23–24). Clearly, this is a major initial signal transduction mechanism for the stimulation of growth and differentiation.

Unfortunately, there is less information about the intracellular pathways affected by the stimulation of a tyrosine kinase; that is, there are few, if any, proteins whose activity is known to be regulated by tyrosine phosphorylation. This has raised doubt as to the importance of the tyrosine kinases in signal transduction. To answer this question, a number of groups have performed site-directed mutagenesis on receptor and membrane-bound tyrosine kinases. In every case tested, mutation of the lysine critical for ATP binding in the kinase region results in an abolition of growth or oncogenic signal transduction (2, 18). Thus, the tyrosine kinase catalytic activity is crucial for growth signaling, but the mechanisms distal to tyrosine phosphorylation are unclear.

Substrate Selectivity

As stated above, there is a degree of sequence conservation in all tyrosine kinase domains. Thus, tyrosine phosphorylation could be restricted to a narrow spectrum of substrates that are similar in all tissues. In this scenario, the multitude of tyrosine kinases would simply reflect tissue diversity, with one or two tyrosine kinases per cell type. The large *src*-like family of kinases does appear to provide multiple tissues with a very similar catalytic domain in the form of at least eight kinases (18, 22). The major sequence differences in this family are confined to the N-terminal domain that functions to anneal the kinase to the inner leaflet of the

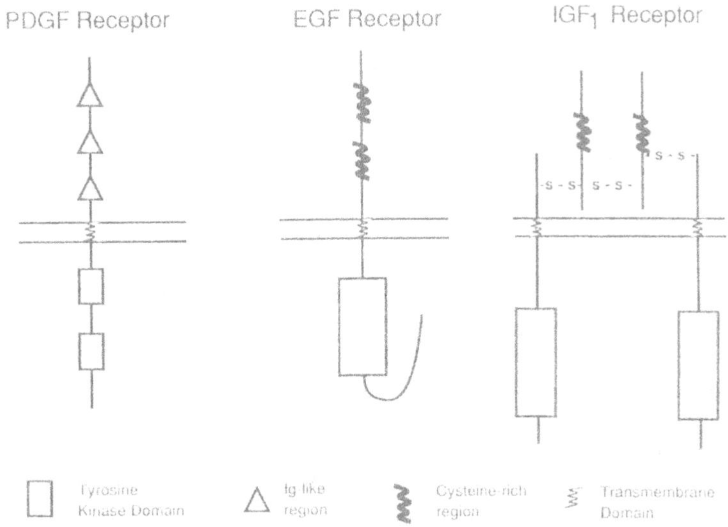

Fig. 1. Growth factor receptor tyrosine-specific protein kinases. A schematic representation shows the three classes of receptor tyrosine kinases. When activated, these receptors phosphorylate the range of substrates required to stimulate Balb/c 3T3 fibroblast growth.

plasma membrane and perhaps to link the kinase to transmembrane molecules with signaling functions. However, between the tyrosine kinase families, there are considerable differences in the 290-amino acid kinase domain. One well-studied example involving growth control of quiescent Balb/c 3T3 fibroblasts suggests that at least three or more tyrosine kinases may have unique functions within a single cell type. Under restricted conditions, three distinct polypeptide growth factors are required to stimulate mitogenesis: PDGF, EGF, and IGF-I (25). Each is needed at defined times, and each binds to and activates a specific receptor with its own unique tyrosine kinase. Selectivity in tyrosine kinase substrates and subsequent downstream signaling mechanisms must exist, or there would be no need to activate all three classes of tyrosine kinases in the single cell type. In fact, these three receptors define three major classes of membrane-spanning tyrosine kinases (Fig. 1).

The first tyrosine kinase-containing receptor sequenced, the EGF receptor, is perhaps the simplest structurally (13). This monomeric protein of 1186 amino acids is almost equally divided into an extracellular ligand-binding domain and an intracellular cytoplasmic domain containing the tyrosine kinase and autophosphorylation sites. These regions are joined in this and all growth factor receptors by a short 20- to 22-amino acid hydrophobic transmembrane-spanning region. The EGF receptor extracellular domain contains two cysteine-rich regions that are characteristic of the other receptors of this class, the *neu* oncogene (26–28) and the recently sequenced ERB B3 (29).

The IGF-I receptor and its relation, the insulin receptor, represent another class of receptor tyrosine kinase (30). Members of this class appear on the surface as a complex structure consisting of subunits that are synthesized from a single gene product. The monomeric form of the IGF-I receptor is processed such that the binding domain is proteolytically cleaved from the domain that has both the transmembrane and tyrosine kinase regions. These two domains remain attached covalently through disulfide linkages. Under most conditions the surface receptor is composed of two of these units that are then disulfide-linked to each other via the extracellular ligand-binding domains. This gives a receptor unit with two ligand-binding sites and two tyrosine kinase domains.

The last major class is typified by the PDGF receptor, whose external ligand-binding domain contains a repeating immunogloublin-like structure rather than cysteine-rich domains (31). The salient feature of the PDGF receptor cytoplasmic domain is peptide sequence that divides the standard 290-amino acid tyrosine kinase domain into two components. Evidence indicates that this kinase insert domain of ~100 amino acids may be important in the identification or binding of selected PDGF receptor substrates (32). Removal of the kinase insert does not abolish tyrosine kinase activity, but does markedly attenuate PDGF-dependent mitogenesis (32–33).

In summary, cells can express and signal differentially through multiple transmembrane receptor tyrosine kinases. Individual cells may utilize several if not many intracellular tyrosine kinases in the course of regulating growth.

EGF Receptor Actions

Figure 2 demonstrates an unactivated EGF receptor within a membrane. The branching carbohydrate chains and the two tightly coiled cysteine-rich regions of the extracellular domain are depicted. Intracellularly, the ATP binding pocket is partly obscured by the C-terminus of the receptor, which is shown folded over into the kinase domain. When EGF binds, a physical signal is transduced, activating the tyrosine kinase. This results in autophosphorylation; that is, the catalytic domain of the receptor adds phosphates to 1–4 tyrosine residues within the C-terminal 100 amino acids of the receptor. The phosphorylated C-terminus is thought to swing away, allowing the catalytic domain access to other substrates that will be phosphorylated on tyrosine residues (see 1 and 2 for review). There continues to be a controversy as to whether the activated EGF receptor functions as a momeric protein or whether EGF binding leads to dimerization of EGF receptors.

In the intact cell, binding of EGF leads to a number of immediate changes in membrane transport and cellular metabolism (Fig. 2). For example, tyrosine phosphorylation increases transport of agents as diverse as amino acids, nucleotides, and sodium, as well as causing rapid changes in cellular pH, glucose, and glycolipid metabolism (1, 6). Each alteration occurs within seconds to minutes after

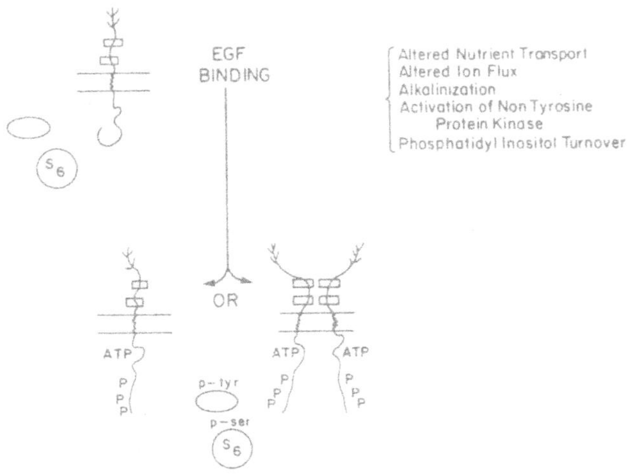

Fig. 2. Activation of the EGF receptor. The binding of EGF alters conformation of the receptor either directly or via a dimerization process. This conformational change stimulates tyrosine kinase activity, resulting in autophosphorylation of C-terminal tyrosines. This leads to the physiologic responses listed on the right. These responses are probably controlled through alteration of the function of tyrosine-phosphorylated proteins. Some of these may be, or may activate, ser/thr protein kinases, such as the one that phosphorylates the S_6 ribosomal protein.

EGF addition and may involve intermediate change in cellular second messengers (Fig. 3). The rapid and general changes in cellular metabolism demonstrate that growth factors, like hormones, are capable of altering differentiated cell function without necessarily triggering cell growth.

Prolonged exposure to growth factors can result in delayed changes in tyrosine phosphorylation, alterations in gene expression, and proliferation. These delayed actions must involve prolonged receptor signaling or a preprogrammed use of nonreceptor tyrosine and, perhaps, serine kinase pathways. The fundamental differences between immediate and prolonged signaling are listed in Figure 4. The well-studied rapid actions begin with ligand binding, which initiates a burst of tyrosine phosphorylation. This occurs when there is a full complement of the receptor on the cell surface. Binding to the surface receptor brings about internalization and, eventually, receptor degradation (1). In liver cells >90% of the receptor is internalized within 10 min, but the internalized receptor remains potentially active as a tyrosine kinase in the endocytotic pathway for the next 60–90 min (34). Thus, some initial EGF actions may be mediated by the internalized EGF receptor tyrosine kinase. For the prolonged actions of EGF, consistent EGF receptor signaling is required over 6–8 h; that is, a constant

Initial Action - Minutes	Prolonged Action - Hours
Burst of Tyrosine Phosphorylation	Cell Growth
Activation of Phospholipases	Mitogenesis
Increase in Intracellular Calcium	Altered Gene Expression
Activation of Protein Kinase C	Delayed tyrosine phosphorylation (cdc 2)
Activation of Ser/Thre Kinases	
Immediate Change in Gene Expression (e.g. c-fos)	

Fig. 3. Effects of EGF on intracellular signaling and processes.

Initial Action - Minutes	Prolonged Action - Hours
a. Begin with 100% of receptor population	a. Requires extracellular ligand
b. Internalization of receptors	b. Involves signalling via a reduced receptor population
c. Degrade Internalized receptors	

Fig. 4. Differences between the status of surface EGF receptor during the rapid and prolonged phases of EGF receptor signaling.

source of extracellular ligand (EGF) and the continued presence of surface receptor are necessary to obtain a proliferation response (1, 6). The surface receptor number is drastically reduced throughout the initial 6–8 h period after EGF addition. This is due to the initial receptor internalization. Thus, a mechanism for retaining some EGF receptor at the cell surface is necessary. Conversely, because EGF's initial signal involves so many second-messenger pathways (Fig. 3 and see below), a short exposure to EGF can have profound cellular effects without causing mitosis.

Substrates for Tyrosine-Specific Protein Kinases

There are now many more tyrosine kinases than there are examples of substrates whose function is regulated by tyrosine phosphorylation. This is due in part to the low abundance of tyrosine phosphate (<1%). The initial studies involved analysis of $^{32}P_i$-labeled cells that had been treated with growth factors or induced to express activated oncogenes. A limited number of substrates, usually phosphorylated at low stoichiometry, were identified subsequently by two-dimensional gel electrophoresis (18). The use of antiphosphotyrosine antibodies for Western blot analyses, immunoprecipitation, and protein purification has allowed more rapid advances in sub-

Identified Substrate	Activity of Substrate
Raf	a Ser/Thr Kinase
MAP Kinase	a Ser/Thr Kinase that phosphorylates the S_6 Kinase
Phospholipase C γ	Production of IP_3 and DAG
GAP	a protein that binds to ras and stimulates ras GTPase activity
PI Kinase	phosphorylates phosphatidyl ionositols with 3' position specificity

Fig. 5. List of known tyrosine-phosphorylated substrates whose phosphorylation might act to alter intracellular signaling.

strate identification (35–37). The identified substrates fall into two general classes. The first are structural proteins, many of which are found within adhesion plaques in cultured cells or in structures of the cytoskeleton. Proteins such as vinculin (38), the fibronectin receptor (39), the calpactins (40), and ezrin (41) are phosphorylated by the unrestrained action of the *src* oncogene or by the addition of growth factors. Their phosphorylation correlates with changes in cell morphology.

Another class of substrates appears to be involved in propagating the downstream signal of the activated growth factor receptor tyrosine kinases (Fig. 5). At least five substrates of this type have been identified in serum or growth factor-stimulated cells, and each has a potential role in signal transduction. However, only for the first example (*raf*) is there more than correlative evidence showing that tyrosine phosphorylation alters the functional capacity of the molecule. The *raf* oncogene is a serine/threonine protein kinase that has been truncated and thus activated, a mutation that results in cell transformation (42, 43). The normal, cellular *raf* is a regulated serine/threonine kinase that can be phosphorylated by the PDGF receptor and activated in a cell-free system by PDGF-dependent tyrosine phosphorylation (44). Phospholipase Cγ (PLCγ) is an isozyme of phosphoinositide-specific phospholipase C (45). Increased tyrosine phosphorylation of this molecule follows EGF (46) and PDGF (47) treatment of cells. Our own lab has demonstrated a tight correlation between tyrosine phosphorylation and dephosphorylation of PLCγ and the initiation and cessation of PLCγ activity in normal and phorbol ester-treated cells (Huckle, et al., unpublished results). The 42-kD MAP kinase, a serine/threonine kinase that is activated by serum stimulation of cells, is phosphorylated on both threonine and tyrosine residues (48). The activity of this enzyme depends upon occupation of both of these phosphorylation sites. The GTP activating protein, GAP (49), is a putative effector for *ras* action and is tyrosine-phosphorylated in PDGF-treated cells (50). Lastly, a tyrosine-phosphorylated phosphotidylinositol kinase with an unusual specificity has been isolated in immunoprecipitates from cells containing v-*src* or polyoma middle T antigen, as well as in cells activated by growth factor receptors (51–53). Whether the activation of this PI kinase by tyrosine phosphorylation stimulates an as yet undefined second-messenger system remains to be determined, but the physical association of phosphorylated PI kinase with the PDGF receptor complex is closely correlated to the mitogenic action of the PDGF receptor.

EGF-Dependent Tyrosine Phosphorylation in WB Cells

Our lab has undertaken detailed studies of activation of the EGF receptor tyrosine kinase in intact WB cells, a continuous line of nontransformed rat liver epithelial cells (54). These cells were cloned from a primary culture of rat hepatocytes in the laboratory of Dr. Joe Grisham. Their surface EGF receptor number is similar to that of rat hepatocytes, ~200,000 per cell. Addition of EGF to intact cells results in rapid clearance of surface EGF receptors (34). As shown in Figure 2, >80% of the receptors are internalized within 5 min. The addition of methylamine to these cultures prevents the acidification of the endocytotic vesicle. This does not change the rate of receptor internalization (Fig. 6), but rather inhibits subsequent degradation of the receptor. This experimental maneuver has allowed us to study the accumulation of activated EGF receptors in the endocytotic compartment. Homogenization of the cells allowed assessment of EGF receptor kinase activity in internalized vesicles at various times after the EGF treatment. Homogenates were incubated with [γ^{32}P]-ATP in vitro. Figure 7 shows that EGF treatment of intact

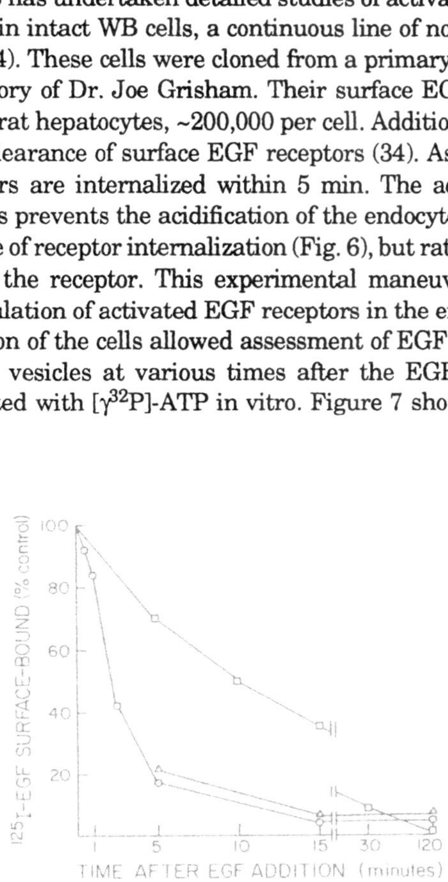

Fig. 6. ^{125}I-EGF surface binding to WB cells. Confluent cultures were incubated with 100-ng/mL EGF for the indicated time periods, acid-washed to remove unlabeled surface-bound EGF, and incubated with ^{125}I-EGF for 2 h at 4°C to determine residual surface-binding activity. Each point represents the average of three determinations minus the nonspecific binding for that sample (standard error <10%). They are expressed as percent of control. (Circles = 37°C; squares = 18°C; triangles = 37°C with 10-mM methylamine.)

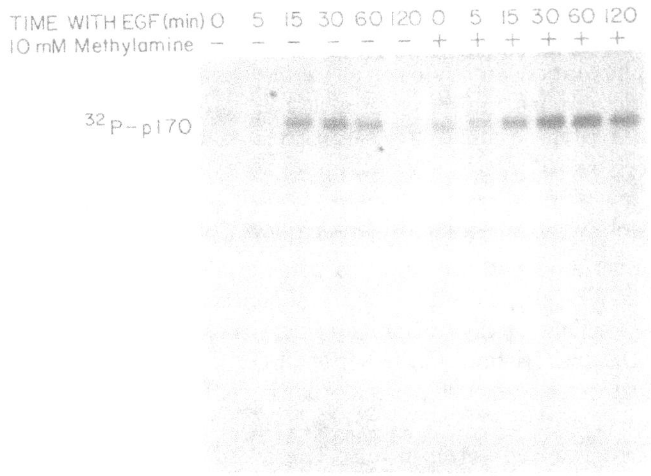

Fig. 7. Autophosphorylation activity of EGF-R in WB cell homogenates. Cultures were incubated with 100-ng/mL EGF at 37°C for the indicated time periods in the absence (left 6 lanes) or presence (right 6 lanes) of 10-mM methylamine. Samples were homogenized and assayed for EGF-R autophosphorylation activity in the presence of 5-μM [(γ-^{32}P)-ATP], 1-mM Mn^{++}, and 0.1% Triton X-100.

cells leads to persistent EGF receptor autophosphorylation activity in internalized receptor preparations. Comparison of Figures 6 and 7 shows that the activated receptor must reside in the internalized vesicles because a vast majority of the receptor has moved to an intracellular site 15–30 min after EGF treatment. Methylamine prevented receptor degradation and allowed the measurement of receptor autophosphorylation 2 h after the addition of EGF to the intact cell. These receptor tyrosine kinase assays have been repeated using an exogenous substrate as the tyrosine phosphate acceptor, with results similar to those obtained with autophosphorylation (55).

We next assessed phosphorylation using antiphosphotyrosine antibodies. We prepared antisera using the method of Kamps and Sefton (37) by linking random copolymers of alanine, glutamine, and phosphotyrosine to KLH. After immunization of rabbits, antisera were obtained and affinity purified using phosphotyrosine-affigel. After elution with phenylphosphate followed by dialysis, highly specific p-Tyr antisera were isolated and used for immunoblots and immunoprecipitation. Figure 8 demonstrates the rapidity with which EGF can stimulate tyrosine phosphorylation in WB cells. Cells were lysed with RIPA buffer (containing 0.1% SDS and other detergents) either before or at various times after the treatment of the intact cell with EGF. Lysates were run on 8% polyacrylamide gels, transferred to nitrocellulose, and immunoblotted as described (3–4). EGF stimulated tyrosine phosphorylation within 5 s at 37°C. The dramatic stimulation of EGF

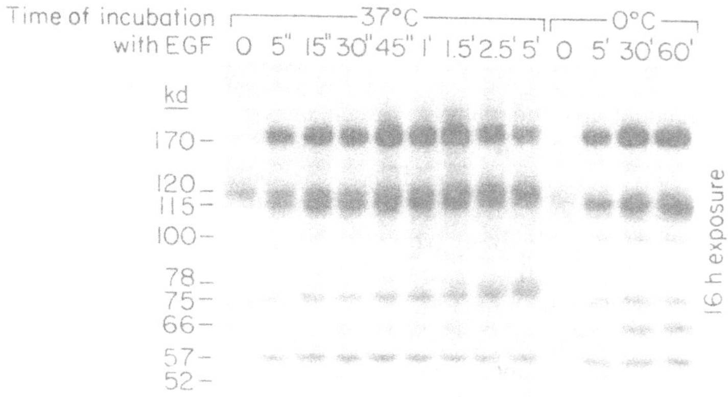

Fig. 8. Effect of temperature and time on EGF-dependent p-Tyr substrates in WB cells. Cells were incubated with 100-ng/mL EGF in culture at 37°C or 0°C for the indicated time periods and analyzed for p-Tyr phosphoproteins by immunoblotting with anti-p-Tyr. Increased immunoidentifiable p-Tyr was initially observed within 5 s in all substrates except p75, p120, and p38 (not seen on this gel). Increases in p-Tyr in these substrates were first observed at ~1 min. In cells treated with EGF at 0°C, increased p-Tyr was seen in all substrates with two exceptions: There was no increase in p-Tyr in p120 and p78 at 0°C.

receptor (p170) phosphorylation is shown and has been confirmed by immuoprecipitation of tyrosine-phosphorylated p170 with anti-EGF receptor antibody.

At least seven other proteins are tyrosine-phosphorylated in an EGF-dependent manner. Intriguingly, there is a subset of tyrosine phosphoproteins whose phosphorylation is stimulated 30–90 s after EGF treatment. These appear as extensions of tyrosine substrates in the 120-kD and 75-kD region and, as seen on lower-percentage gels, a band in the 38-kD region. EGF-dependent tyrosine phosphorylation proceeded even when intact cells were treated at 0°C, albeit at a somewhat slower rate. However, when tyrosine phosphorylation is performed at 0°C, there is no evidence of the tyrosine phosphorylation of the 120-, 78-, or 38-kD substrates noted above. This leads us to postulate that the EGF receptor tyrosine phosphorylation can be divided into two phases: (*a*) rapid phosphorylation of membrane and perimembrane substrates in close proximity to the EGF receptor (these substrates can be phosphorylated at 0°C); and (*b*) phosphorylation of another set of substrates >30 s after EGF treatment and only at temperatures >4°C.

There are at least four possible explanations for this distinction among substrates. The first is that the EGF receptor tyrosine kinase is internalized in the endocytotic vesicles and is capable of phosphorylating new substrates as it moves through the endocytotic pathway. The second possibility is that the shape changes attending EGF action may bring certain proteins into proximity with surface EGF receptors; in other words, it is the substrates that move to the receptor. The third possibility is that the EGF receptor tyrosine kinase activates another tyrosine kinase. These three explanations are not mutually exclusive. Longer-term studies of EGF-dependent tyrosine phosphorylation showed an additional substrate that appears 5–15 min after EGF treatment in the 78–80-kD region; this is another candidate for an internalization-specific receptor substrate. We also have preliminary evidence indicating that EGF action in WB cells may regulate tyrosine phosphorylation indirectly through other kinases. A last possibility that cannot be excluded is that some of the tyrosine substrates (e.g., p75) are subject to delayed phosphorylation by serine/threonine kinases. This would not change the amount of p-Tyr on the immunoblots, but might result in the shift of apparent molecular weight of a tyrosine phosphoprotein. This secondary serine phosphorylation that changes the mobility of the phosphoprotein may not occur at 0°C.

In summary, while a number of issues must be resolved, it is clear that a subset of substrates are phosphorylated within 5 s, while additional substrates are phosphorylated only with a more prolonged activation of the EGF receptor. With the exception of the EGF receptor and PLCγ, we have not yet identified the tyrosine substrates in WB cells. Preliminary experiments in WB cells indicate that p120 is not GAP, nor is the p75 protein the proto-oncogene *raf*.

Regulation of EGF Receptor Biosynthesis

The initial sensitivity of a cell to the action of EGF is determined by the number and affinity of EGF receptors that it displays on the cell surface. In addition to the initial actions, mitogenesis requires prolonged growth factor action at the surface of the cell and must therefore require a supply of EGF receptor on the surface. This could occur either by a mechanism that prevents internalization of the last 10% of receptors remaining at the surface after the initial (10 min) internalization or by a mechanism that replenishes EGF receptor to the surface after the initial internalization. While studying the biosynthesis of EGF receptor in the rat liver epithelial cells, we demonstrated that EGF stimulated EGF receptor biosynthesis. The studies began by prelabeling cells with [^{35}S]methionine and adding EGF. As expected, EGF caused degradation of prelabeled EGF receptor within 30–60 min. When the experiment was done in the opposite manner—that is, EGF was added, then EGF receptor biosynthesis was followed by addition of [^{35}S]methionine—EGF increased EGF receptor protein synthesis (56–57). Figure 9A shows a dose-dependent increase in EGF receptor biosynthesis measured 4 h after EGF treatment of WB cells. The time-course of EGF-induced EGF receptor synthesis is shown in Figure 9B. EGF was added for 1, 2, and 4 h, and receptor synthesis was studied during the last hour. With the short labeling period, both the nascent p160 form of the receptor and the mature p170 form of the receptor are seen. By 2 h, increased synthesis of p160 is seen. It is the initial burst of EGF action that causes this phenomenon, as withdrawal of EGF after 15 min did not block the subsequent increase in receptor synthesis at 2–4 h. Additionally, inhibition of receptor degradation with methylamine does not block the signal that increases receptor synthesis. Hence, it is the

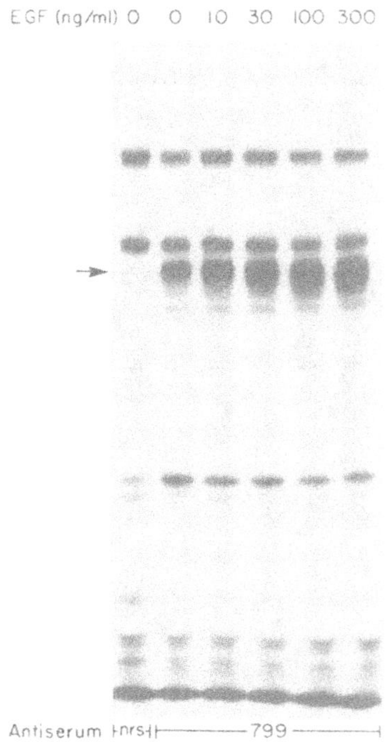

Fig. 9A. EGF-stimulated EGF receptor synthesis: Dose. Cells were incubated for 4 h with [^{35}S]methionine with or without the indicated EGF concentrations (ng/mL). Immunoprecipitation was performed with normal rabbit sera (nrs) (lane 1) or antisera 799 (lanes 2–6). This was followed by 6% polyacrylamide electrophoresis fluorography and autoradiography. The arrow denotes p170.

receptor's initial signal, not receptor degradation, that triggers the increased receptor synthesis.

The mechanism of EGF-stimulated receptor synthesis involves accumulation of EGF receptor mRNA. Northern blots of total and poly A-selected mRNA isolated from WB cells at various times after EGF treatment have demonstrated that EGF increases receptor mRNA levels (57). During the course of these experiments, it was also demonstrated that EGF receptor mRNA accumulation could be stimulated by agents that activate phospholipase C and protein kinase C. Figure 10 shows that epinephrine, EGF, TPA, and [Arg^8]vasopressin stimulated the accumulation of the major EGF receptor transcript of 9.6 kb. Forskolin, which stimulates cAMP production, did not increase EGF receptor mRNA. Figure 10 also shows the time course of EGF receptor mRNA accumulation in epinephrine-treated cells;

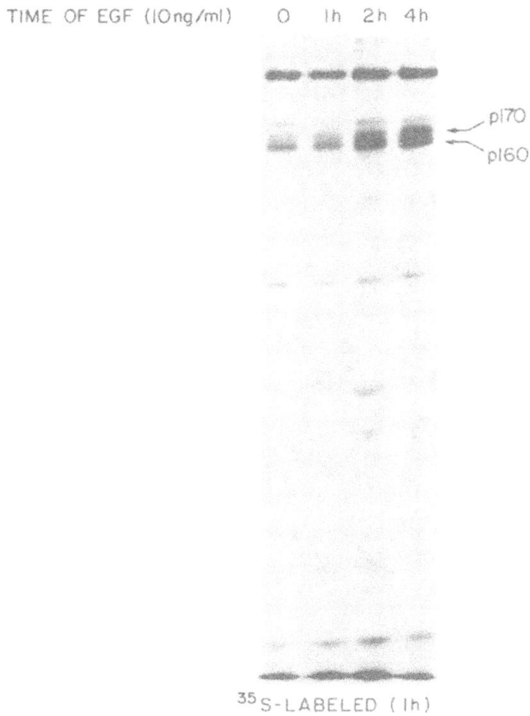

Fig. 9B. EGF-stimulated EGF receptor synthesis: Time-Course. Time-course of EGF-induced EGF receptor synthesis. Confluent WB cells were incubated in 2% methionine-containing medium for 4 h. EGF (10 ng/mL) was added to the cells for the indicated time (0 represents a 4-h incubation without EGF); 50 µCi of [^{35}S]methionine was present for the last 1 h in all samples. Immunoprecipitation was followed by 8% sodium dodecyl sulfate polyacrylamide gel electrophoresis and fluorography. The short labeling time allows the detection of both the nascent (p160) and mature (p170) form of the receptor.

the accumulation began between 1–2 h and was maximal at ~3 h. The time-courses of EGF and epinephrine action are displayed using densitometric tracings (Fig. 11).

A common mechanism for the effect of EGF, TPA, and hormones could have been activation of protein kinase C. To test this hypothesis, WB cells were preincubated with 10-µM TPA for 18 h, a treatment that depletes >95% of the cellular-protein kinase C activity (PKC) (57). The depletion abolishes the effect of TPA on EGF receptor biosynthesis, but does not alter the effect of EGF. In the aggregate, the effects of epinephrine and angiotensin are reduced by ~50%. Thus, EGF stimulates EGF receptor mRNA accumulation by a mechanism that is independent of PKC.

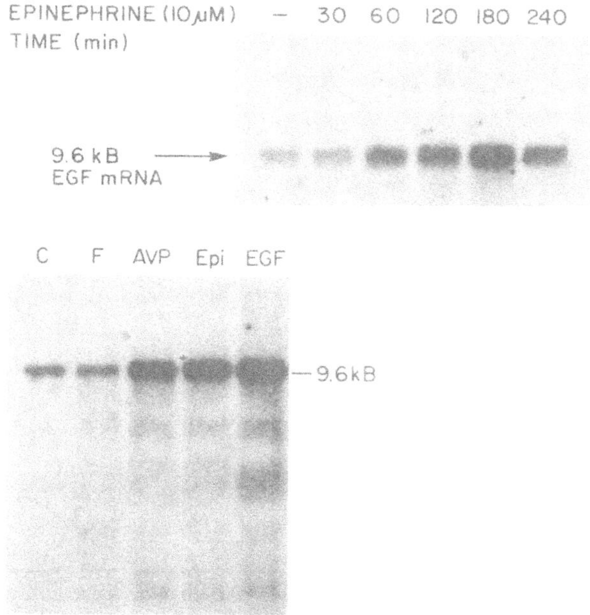

Fig. 10. Effect of epinephrine, [Arg[8]]vasopressin, EGF, and forskolin on EGF receptor mRNA. In the top panel, total RNA was isolated from WB cells at various times after treatment with epinephrine (10 μM) and was subjected to 1%-agarose gel electrophoresis. After transfer to nitrocellulose, hybridization was performed with a nick-translated 1.8-kb cDNA from the coding region of the rat EGF receptor. In the bottom panel, WB cells were treated for 3 h with 30-μM forskolin (F), 1-μM [Arg[8]]vasopressin (AVP), 10-μM epinephrine (Epi), and 100-ng/mL EGF, or vehicle (C). Total RNA was isolated and analyzed as described above.

This unique mechanism of EGF is important for two reasons. First, this action of EGF on its receptor is a paradigm for EGF alteration of gene expression. EGF-directed EGF receptor biosynthesis, which has been seen in several other cell types (58–59), may provide insight into how growth factors alter gene expression. Second, EGF-stimulated EGF receptor synthesis may assure that new surface receptor is supplied over the 6–8 h needed to propagate the mitogenic action of EGF. This effect of EGF is moderately rapid; that is, the initial burst of tyrosine phosphorylation appears to send a preset signal, resulting in peak EGF receptor RNA accumulation within 3 h; receptor synthesis approaches basal by 6–8 h. When EGF receptor mRNA levels are monitored in populations of cycling WB cells, it appears that EGF receptor synthesis decreases as cells leave quiescence and begin to grow rapidly. Thus, the confluent quiescent WB cells are programmed to respond

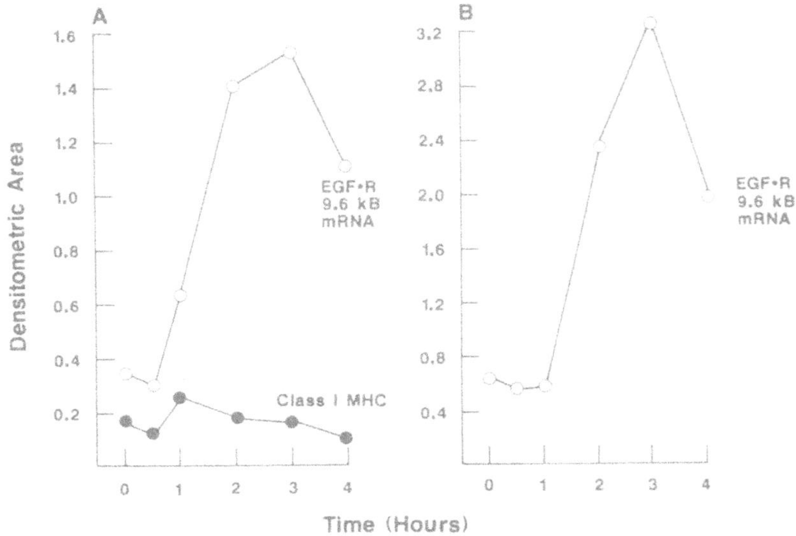

Fig. 11. Time-course of EGF receptor (EGF-R) 9.6-kb mRNA accumulation after stimulation with epinephrine (10 μM) and EGF (100 ng/mL). Total RNA was isolated and Northern blots performed. Hybridization was performed with EGF receptor cDNA. Densitometric scans were performed, and the integrated areas were graphed as a function of time and epinephrine (A) and EGF (B). The RNA blot from epinephrine-treated cells was stripped by washing in boiling 0.1 × SSC, 0.1% sodium dodecyl sulfate and was reprobed with an MHC class 1 sequence.

to the initial burst of EGF action by replenishing the surface receptor complement for a time sufficient to allow a mitogenic signal to be induced (6–8 h). Once the cell cycle has been initiated, receptor biosynthesis appears to decrease. A similar decrease in receptor mRNA is also seen when hepatocytes enter the S-phase in vivo during liver regeneration (Marts, et al., unpublished results).

CONCLUSION

The EGF and other growth factor receptors are self-contained signaling units that are capable of binding ligand with their external domains and transducing a signal via the intrinsic tyrosine kinase of their cytoplasmic domains. The tyrosine kinase is involved in signal transduction, but the mechanism by which the signal is propagated after the initial activation of the kinase is unknown. The initial signal alters aspects of cellular metabolism and activates a number of second-messenger systems. Many of these same messenger systems are activated in other circumstances by classical hormones. Thus, growth factors may, with short exposure, regulate normal metabolic functions of differentiated tissues. It is only when growth factors act for prolonged periods, in concert with other hormones and growth factors, that they stimulate mitogenesis. This requirement for prolonged growth factor action is not well understood mechanistically. In the reproductive tract there is an enormous proliferative potential that is regularly mediated, at least in females, by alterations in classical hormones. The circuits by which the classical hormones influence growth factor production, release, and signaling via growth factor receptors are important topics for further research.

REFERENCES

1. Carpenter G. Receptors for epidermal growth factor and other polypeptide mitogens. Annu Rev Biochem 1987;56:881-914.
2. Yarden Y, Ullrich A. Growth factor receptor tyrosine kinases. Annu Rev Biochem 1988;57:443-78.
3. Samelson LE, Patel MD, Weissman AM, Harford JB, Klausner RD. Antigen activation of murine T cells induces tyrosine phosphorylation of a polypeptide associated with the T cell antigen receptor. Cell 1986;46:1083-90.
4. Katagiri T, Ting JP-Y, Dy R, Prokop C, Cohen P, Earp HS. Tyrosine phosphorylation of a c-Src-like protein is increased in membranes of $CD4^-$ $CD8^-$ T lymphocytes from lpr/lpr mice. Mol Cell Biol 1989;9:4914-22.
5. Cohen S. Isolation of a mouse submaxillary gland protein accelerating incisor eruption and eyelid opening in the newborn animal. J Biol Chem 1962;237:1555-62.
6. Carpenter GE, Cohen S. Epidermal growth factor. Annu Rev Biochem 1979;48: 193-216.
7. Bishop JM. The molecular genetics of cancer. Science 1987;235:305-11.
8. Heldin CH, Westermark B. Growth factors: Mechanism of action and relation to oncogenes. Cell 1984;37:9-20.
9. Devare SG, Reddy EP, Law JD, Robbins KC, Aaronson SA. Nucleotide sequence of the simian virus genome: Demonstration that its acquired cellular sequences encode the transforming gene product $p28^{sis}$. Proc Natl Acad Sci 1983;80:731-5.
10. Doolittle RF, Hunkapiller MW, Hood LE, et al. Simian sarcoma virus oncogene, v-sis, is derived from the gene (or genes) encoding a platelet-derived growth factor. Science 1983;221:275-7.
11. Waterfield MD, Scrace T, Whittle N, et al. Platelet-derived growth factor is structurally related to the putative transforming protein $p28^{sis}$ of simian sarcoma virus. Nature 1983;304:35-9.
12. Downward J, Yarden Y, Mayes E, et al. Close similarity of epidermal growth factor receptor and v-erb-B oncogene protein sequences. Nature 1984;307:521-7.
13. Ullrich A, Coussens L, Hayflick JS, et al. Human epidermal growth factor receptor cDNA sequence and aberrant expression of the amplified gene in A431 epidermoid carcinoma cells. Nature 1984;309:418-25.
14. Sefton BM, Hunter T, Beemon K, Eckhart W. Evidence that the phosphorylation of tyrosine is essential for cellular transformation by Rous sarcoma virus. Cell 1980; 20:807-16.
15. Eckhart W, Hutchinson MA, Hunter T. An activity phosphorylating tyrosine in polyoma T antigen immunoprecipitates. Cell 1979;18:925-33.
16. Witte ON, Dasgupta A, Baltimore D. Abelson murine leukaemia virus protein is

phosphorylated in vitro to form phosphotyrosine. Nature 1980;283:826-8.
17. Collett MS, Purchio AF, Erikson RL. Avian sarcoma virus-transforming protein, pp60src shows protein kinase activity specific for tyrosine. Nature 1980;285:167-9.
18. Hunter T, Cooper JA. Protein-tyrosine kinases. Annu Rev Biochem 1985;54:897-930.
19. Carpenter G, King L Jr, Cohen S. Epidermal growth factor stimulates phosphorylation in membrane preparations in vitro. Nature 1978;276:409-10.
20. Ushiro H, Cohen S. Identification of phosphotyrosine as a product of epidermal growth factor-activated protein kinase in A-431 cell membranes. J Biol Chem 1980; 255:8363-5.
21. Kamps MP, Buss JE, Sefton BM. Mutation of NH2-terminal glycine of $p60^{v\text{-}src}$ prevents both mytistylation and morphological transformation. Proc Natl Acad Sci 1985;82:4625-8.
22. Hanks SK, Quinn AM, Hunter T. The protein kinase family: Conserved features and deduced phylogeny of the catalytic domains. Science 1988;241:42-52.
23. Hafen E, Basler K, Edstroem JE, Rubin GM. Seven-less, a cell-specific homeotic gene of *Drosophila*, encodes a putative transmembrane receptor with a tyrosine kinase domain. Science 1987;236:55-63.
24. Sprenger F, Stevens LM, Nüsslein-Volhard C. The *Drosophilia* gene torso encodes a putative receptor tyrosine kinase. Nature 1989;338:478-83.
25. O'Keefe EJ, Pledger WJ. A model of cell cycle control: Sequential events regulated by growth factors. Mol Cell Endocrinol 1983;31:167-86.
26. Schechter AL, Stern DF, Vaidyanathan L, et al. The neu oncogene: An erb-B-related gene encoding a 185,000-Mr tumour antigen. Nature 1984;312:513-6.
27. Bargmann CI, Hung MC, Weinberg RA. The neu oncogene encoding an epidermal growth factor receptor-related protein. Nature 1986;319:226-30.
28. King CR, Kraus MH, Aaronson SA. Amplification of a novel v-erb B-related gene in a human mammary carcinoma. Science 1985;229:974-8.
29. Kraus MH, Issing W, Miki T, Aaronson SA. Isolation and characterization of ERB B3, a third member of the ERB B/epidermal growth factor receptor family: Evidence for overexpression in a subset of human mammary tumors. Proc Natl Acad Sci 1989; 86:9193-7.
30. Ullrich A, Gray A, Tam AW, et al. Insulin-like growth factor I receptor primary structure: Comparison with insulin receptor suggests structural determinants that define functional specificity. EMBO J 1986;5:2503-12.
31. Yarden Y, Escobedo JA, Kuang WJ, et al. Structure of the receptor for platelet-derived growth factor helps define a family of closely related growth factor receptors. Nature 1986;323:226-32.
32. Escobedo JA, Williams LT. A PDGF receptor domain essential for mitogenesis but not for many other responses to PDGF. Nature 1988;335:85.
33. Williams LT. Signal transduction by the platelet-derived growth factor receptor. Science 1989;243:1564-70.
34. McCune BK, Earp HS. The epidermal growth factor receptor tyrosine kinase in liver epithelial cells. J Biol Chem 1989;264:15501-7.
35. Wang JYJ. Isolation of antibodies for phosphotyrosine by immunization with a v-abl oncogene-encoded protein. Mol and Cell Biol 1985;5:3640-3.
36. Morla AO, Draetta G, Beach D, Wang YJ. Reversible tyrosine phosphorylation of cdc2: Dephosphorylation accompanies activation during entry into mitosis. Cell 1989;58: 193-203.

37. Kamps MP, Sefton BM. Identification of multiple novel polypeptide substrates of the v-src, v-yes, v-fps, v-ros, and v-erb-B oncogenic tyrosine protein kinases utilizing antisera against phosphotyrosine. Oncogene 1988;2:305-15.
38. Sefton BM, Hunter T, Ball EH, Singer SJ. Vinculin is a cytoskeletal target of the transforming protein of Rous sarcoma virus. Cell 1981;24:165-74.
39. Hirst R, Horwitz A, Buck C, Rohrschneider L. Phosphorylation of the fibronectin receptor complex in cells transformed by oncogenes that encode tyrosine kinases. Proc Natl Acad Sci 1986;83:6470-4.
40. Glenney J. Two related but distinct forms of the M_r 36,000 tyrosine kinase substrate (calpactin) that interact with phospholipid and actin in a Ca^{2+}-dependent manner. Proc Natl Acad Sci 1986;83:4258-62.
41. Gould KL, Bretscher A, Esch FS, Hunter T. cDNA cloning and sequencing of the protein-tyrosine kinase substrate, ezrin, reveals homology to band 4.1. EMBO J 1989; 8:4133-42.
42. Rapp UR, Cleveland JL, Storm SM, Beck TW, Huleihel M. Transformation by raf and myc oncogenes. In: Aaronson SA, ed. Oncogenes and cancer. Tokyo: Japan Sci Soc Press, 1987b:55-74.
43. Morrison DK, Kaplan DR, Roberts TM. Signal transduction from membrane to cytoplasm: Growth factors and membrane-bound oncogene products increase Raf-1 phosphorylation and associated protein kinase activity. Proc Natl Acad Sci 1988; 85:8855-9.
44. Morrison DK, Kaplan DR, Escobedo JA, Rapp UR, Roberts TM, Williams LT. Direct activation of the serine/threonine kinase activity of Raf-1 through tyrosine phosphorylation by the PDGF β-receptor. Cell 1989;58:649-57.
45. Rhee SG, Suh PG, Ryu SH, Lee SY. Studies of inositol phospholipid-specific phospholipase C. Science 1989;244:546-50.
46. Wahl MI, Nishibe S, Suh PG, Rhee SG, Carpenter G. Epidermal growth factor stimulates tyrosine phosphorylation of phospholipase C-II independently of receptor internalization and extracellular calcium. Proc Natl Acad Sci 1989;86:1568-72.
47. Meisenhelder J, Suh PG, Rhee SG, Hunter T. Phospholipase C-γ is a substrate for the PDGF and EGF receptor protein-tyrosine kinases in vivo and in vitro. Cell 1989;57: 1109-22.
48. Anderson NG, Maller JL, Tonks NK, Sturgill TW. Requirement for integration of signals from two distinct phosphorylation pathways for activation of MAP kinase. Nature 1990;343:651-3.
49. Ellis C, Moran M, McCormick F, Pawson T. Phosphorylation of GAP and GAP-associated proteins by transforming and mitogenic tyrosine kinases. Nature 1990; 343:377-81.
50. Molloy CJ, Bottaro DP, Fleming TP, Marshall MS, Gibbs JB, Aaronson SA. PDGF induction of tyrosine phosphorylation of GTPase activating protein. Nature 1989; 342:711-4.
51. Courtneidge SA, Heber A. An 81 kD protein complexed with middle T antigen and $pp60^{c\text{-}src}$: A possible phosphatidylinositol kinase. Cell 1987;50:1031-7.
52. Kaplan DR, Whitman M, Schaffhausen B, et al. Common elements in growth factor stimulation and oncogenic transformation: 85 kD phosphoprotein and phosphatidylinositol kinase activity. Cell 1987;50:1021-9.
53. Coughlin SR, Escobedo JA, Williams LT. Role of phosphatidylinositol kinase in PDGF receptor signal transduction. Science 1989;243:1191-4.

54. Tsao MS, Smith JD, Nelson KG, Grisham JW. A diploid epithelial cell line from normal adult rat liver with phenotypic properties of "oval" cells. Exp Cell Res 1984; 154:38-52.
55. McCune BK, Earp HS. EGF-dependent suppression of epidermal growth factor receptor autophosphorylation in liver epithelial cells is transient and is not due to activation of protein kinase C. J Biol Chem (in press).
56. Earp HS, Austin KS, Blaisdell J, et al. Epidermal growth factor (EGF) stimulates EGF receptor synthesis. J Biol Chem 1986;261:4777-80.
57. Earp HS, Hepler JR, Petch LA, et al. Epidermal growth factor (EGF) and hormones stimulate phosphoinositide hydrolysis and increase EGF receptor protein synthesis and mRNA levels in rat liver epithelial cells. J Biol Chem 1988;263:13868-74.
58. Clark AJL, Ishii S, Richert N, Merlino GT, Pastan I. Epidermal growth factor regulates the expression of its own receptor. Proc Natl Acad Sci 1985;82:8374-8.
59. Kudlow JE, Cheung CYM, Bjorge JD. Epidermal growth factor stimulates the synthesis of its own receptor in a human breast cancer cell line. J Biol Chem 1986;261: 4134-8

2

Transforming Growth Factor β: A Multifunctional Regulatory Peptide with Actions in the Reproductive System

Kathleen C. Flanders, Belinda A. Marascalco, Anita B. Roberts, and Michael B. Sporn

Laboratory of Chemoprevention, National Cancer Institute, Bethesda, Maryland

Transforming growth factor β (TGFβ) was first identified as a factor that could induce normal rat kidney (NRK) fibroblasts to form colonies in soft agar in the presence of epidermal growth factor (EGF) (1). Even though TGFβ has the ability to act in this classical assay for transformation, we now know that it is also a mediator of normal cellular physiology and has especially important actions in the processes of embryonic development, tissue remodeling, and wound healing (2). Almost all cells in culture synthesize TGFβ and have TGFβ receptors (3), and immunoreactive TGFβ has been found in a number of embryonic and adult murine tissues (4–5). These data suggest that TGFβ has widespread biological actions. In this chapter we review the chemistry and biology of the TGFβ family, with special emphasis on some of the biological actions of TGFβ that are most likely to be important in the function of the reproductive tract. A brief overview of the actions of TGFβ on gonadal cell types is given here; more details can be found in other chapters of this volume.

TGFβ SUPERFAMILY

Structurally, the active form of the first TGFβ cloned and sequenced (6) is a disulfide-linked homodimer with a molecular weight of 25 kD in the unreduced form. Each monomer consists of 112 amino acids. This TGFβ ($TGF\beta_1$) is synthesized as a 390-amino acid precursor, of which the mature peptide comprises the carboxyl terminal sequence. In the past few years, four additional TGFβs have been cloned: $TGF\beta_2$ was cloned from simian, human, and chicken libraries (7–9); $TGF\beta_3$, from human, pig, mouse, and chicken libraries (10–13); $TGF\beta_4$, from a chicken chondrocyte library (14); and $TGF\beta_5$, from a frog oocyte library (15). All TGFβs have been purified with the exception of $TGF\beta_4$. The TGFβ monomers are all 70%–80%

identical, and there is complete conservation of the 9 cysteine residues in each TGFβ monomer. As with $TGF\beta_1$, each monomer is 112–114-amino acids long and occupies the carboxyl terminal sequence of its precursor. The proregions of the TGFβ precursors are less highly conserved between isoforms than are the mature regions. The sequences for each of the mature TGFβ isoforms are 98%–100% conserved between species; this high degree of conservation between species suggests that the TGFβs have important biological roles.

TGFβs 1,2,3, and 5 crossreact with the same receptor system, and in many in vitro assays, such as induction of growth of NRK cells in soft agar and inhibition of epithelial cell growth, they behave identically (16–17). Some differences in activity are beginning to be found: $TGF\beta_1$ is much more potent in inhibiting the growth of endothelial cells than is $TGF\beta_2$ (18), while $TGF\beta_3$ is most active in assays measuring mesoderm induction in amphibians (16). In vivo, there appears to be selective expression of TGFβ isoforms. For example, mouse, chick, and frog embryos express predominantly TGFβs 1, 3, and 5, respectively. The TGFβ isoforms also seem to be regulated differently: Primary keratinocytes treated with retinoic acid selectively increase expression of $TGF\beta_2$ with little change in expression of $TGF\beta_1$(19). Areas of active investigation are the significance of the expression of several TGFβ isoforms in a tissue, along with discerning the in vivo actions of different TGFβ isoforms, and the control of expression of each isoform.

In addition to five distinct TGFβs, there are now many peptides that belong to the TGFβ supergene family by virtue of their 30% to 40% amino acid homologies to the TGFβs and conservation of 7 of the 9 cysteine residues of TGFβ. The peptides are encoded as larger precursors, and the family resemblance is limited to the C-terminus of the larger precursor corresponding to the processed mature TGFβ. The C-terminus of each peptide ends in the sequence Cys-X-Cys-X. These peptides and their biological activities are listed in Table 1.

TGFβs IN EMBRYOGENESIS

A common feature of the biology of the peptides listed in Table 1 is their ability to regulate developmental processes. TGFβs themselves seem to play a role in development. In amphibians $TGF\beta_3$ is able to induce the formation of mesoderm from ectoderm in vitro (16), while $TGF\beta_1$ augments the ability of fibroblast growth factor (FGF) to induce mesoderm (27). Immunoreactive $TGF\beta_5$ has been localized to ectoderm and mesoderm in *Xenopus* embryos through the neurula stage (K. Flanders, unpublished). TGFβs are also present throughout development of the mouse embryo. $TGF\beta_1$ mRNA and protein appear after fertilization (28) and continue to be expressed throughout embryogenesis (4, 29). In mouse embryos of 11–18 days' gestation, $TGF\beta_1$ protein is often localized to areas of critical epithelial-mesenchymal interactions (4), as in the mesenchyme of the developing hair follicles, teeth, and submandibular gland. Regions of tissue remodeling, such as in mesenchyme underlying developing digits, heart valves, and palate, are also intensely stained.

$TGF\beta_1$ is also expressed both spatially and temporally in the developing embryo. For example, the pattern of TGFβ staining in the developing somites dem-

Table 1. Members of the TGFβ gene family.

Peptide	Homology (%)	Function	Reference
TGFβ	100	Multifunctionally regulates cell growth and differentiation	2
Inhibins	28–38	Inhibit secretion of FSH by pituitary cells	20
Activins	33–38	Stimulate secretion of FSH by pituitary cells	21
MIS	32	Induces regression of mullerian ducts in male embryos	22
DPP-C	36	Establishes dorsal-ventral pattern in *Drosophila* embryos	23
Vg_1	38	Functions in mesoderm induction in frog development	24
Vgr_1	34	May be involved in mammalian tissue differentiation	25
BMP	36	Induces formation of cartilage in vivo	26

Note: MIS = mullerian inhibitory substance; DPP-C = decapentaplegic gene complex; BMP = bone morphogenetic protein.

onstrates that $TGF\beta_1$ contributes to the segmentation of the axial skeleton: Staining is uniform throughout the primitive somite, but subsequently localizes in the sclerotome and dermatome as development progresses and, finally, in the area defining the centrum of the future definitive vertebrae. mRNA for TGFβs 2 and 3 are also expressed during mouse embryogenesis, and localization of these TGFβ proteins is in progress. In some cells, such as cardiac myocytes and chondrocytes, staining for these TGFβs are similar, while in other cell types, such as neurons and glia, TGFβs 2 and 3 are predominantly expressed (K. Flanders, unpublished). The localization of TGFβs in chondrocytes, osteocytes, and osteoblasts in the mouse embryo correlates with the effects of TGFβs on these cells in vitro. Fetal bovine osteoblasts (30) and mesenchymal precursors of chondrocytes (31) are stimulated to grow by TGFβ. The presence of TGFβs during development, as well as its effects in control of cell migration, cell growth and differentiation, angiogenesis, and function and regulation of extracellular matrix, implicate its involvement in embryogenesis. These biological actions of TGFβ are discussed next.

BIOLOGICAL ACTIONS OF TGFβ

TGFβ is a prototype of a multifunctional growth factor (32). TGFβ not only affects a variety of cellular functions, but its action on a target cell can depend on the cell type and its state of differentiation, growth conditions, and other growth factors present. The mechanism by which TGFβ mediates its diverse biological effects is not known. Since most cells in culture both produce TGFβ and possess TGFβ receptors, the ability to restrict its biological actions to appropriate tissues at appropriate times is important. TGFβ is secreted by platelets and cells in culture in a biologically inactive, latent form that cannot bind to the TGFβ receptor (33). The latent form of TGFβ consists of the active TGFβ dimer noncovalently associated with a dimer of the proregion of the precursor. In some cases a third protein is found in the complex. In vitro, latent TGFβ can be activated by treatment with heat, acid, or SDS. The physiological mechanism of activation has not been demonstrated. Limiting the activation of latent TGFβ to particular cell types under specific conditions may be a mechanism to regulate the varied actions of TGFβ, which are discussed below.

Effects of TGFβ on Cell Growth and Differentiation

Treatment of mesenchymally derived cells in culture with TGFβ generally stimulates growth, although growth effects can be modified by the presence of other growth factors (2). The effects of TGFβ on the differentiation of mesenchymal cells may result from the increased production of extracellular matrix proteins that occurs when cells are treated with TGFβ (see below). Thus, treatment of chondroblasts with TGFβ inhibits their growth and alters the state of differentiation of the cells by changing the composition of the extracellular matrix proteins that the cells secrete (34).

TGFβ inhibits the growth of most epithelial cells in culture, including keratinocytes; renal tubular cells; and bronchial, tracheal, intestinal, and prostate epithelial cells (2, references). Even in vivo, slow-release implants containing TGFβ have been shown to inhibit mammary growth and morphogenesis, resulting in complete inhibition of end bud formation (35). Human mesothelial cells (36) and rat uterine and vaginal epithelial cells (37) appear to be exceptions in that their growth is stimulated by treatment with TGFβ. Since inhibition of epithelial cells is often accompanied by their terminal differentiation, TGFβ promotes differentiated function in tracheal, bronchial, and intestinal epithelial cells. With respect to effects on differentiation, TGFβ is again a bifunctional agent: While inducing differentiation of the cells mentioned above, TGFβ inhibits differentiation of adipocytes and myocytes and modulates the differentiated state of granulosa and Leydig cells (see below).

Effects of TGFβ on Angiogenesis

TGFβ appears to be angiogenic in several in vivo assay systems (2), and neovascularization is necessary in processes where TGFβ plays a prominent role, such as embryogenesis and wound healing. In the reproductive tract, formation of new blood vessels is especially important in functioning of the placenta and in uterine cycling. Because TGFβ has effects on so many cell types, it is not known if neovascularization results from a direct effect of TGFβ on endothelial cells or from indirect stimulation of endothelial cells by other growth factors that are induced by TGFβ. Indeed, $TGF\beta_1$ is a potent inhibitor of endothelial cell growth in monolayer culture, although $TGF\beta_2$ is much less active (18). Thus, a switch in expression of

TGFβ isoforms may regulate the effects of TGFβ on angiogenesis. When endothelial cells are cultured three-dimensionally in collagen gels, $TGF\beta_1$ does not inhibit proliferation, but promotes organization of endothelial cells into tubelike structures (38). This indicates that the induction of angiogenesis by TGFβ could be the consequence of an organizational effect, rather than a growth effect, on endothelial cells.

Effect of TGFβ on Production of Extracellular Matrix

The ability of TGFβ to increase the accumulation and response of cells to extracellular matrix proteins is, perhaps, the most well characterized biological action of TGFβ. Figure 1 summarizes the mechanisms by which TGFβ mediates these effects. TGFβ enchances accumulation both by increasing synthesis of the matrix proteins by cells and by inhibiting degradation of these proteins by proteases. TGFβ is chemotactic for fibroblasts (39) and also causes increased synthesis of fibronectin and type I collagen (40). Other more specialized matrix proteins, such as types III, IV, and V collagens, thrombospondin, osteopontin, tenascin, elastin, osteonectin, and chondritin/dermatan sulfate proteoglycans are also induced by TGFβ (2, references). This enhancement of synthesis results from both activation of promoter elements of some matrix proteins (41), as well as stabilization of mRNAs (42).

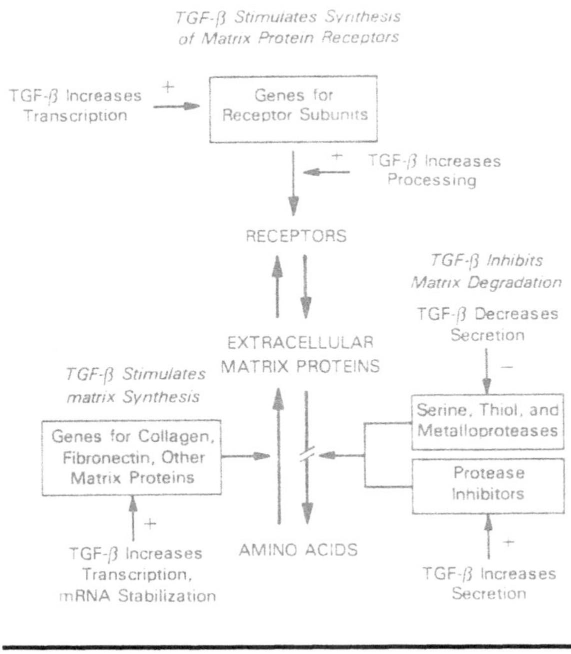

Fig. 1. Mechanisms by which TGFβ increases accumulation of extracellular matrix proteins and integrins. (+ = effects that are increased by TGFβ; − = effects decreased by TGFβ.) Details are provided in the text.

TGFβ also alters mRNA levels of genes that regulate matrix degradation. TGFβ decreases synthesis and secretion of proteases, such as plasminogen activator, collagenase, major excreted protein, elastase, and transin, while increasing synthesis and secretion of protease inhibitors, such as plasminogen activator inhibitor and tissue inhibitor of metalloproteases (2). Thus, by these two mechanisms TGFβ inhibits degradation of matrix proteins. TGFβ also increases interaction of cells with the increased levels of extracellular matrix proteins by increasing expression of cell membrane receptors for cell adhesion proteins (43). These receptors, called integrins, are transmembrane links between the extracellular matrix and cytoskeletal elements of the cell. TGFβ increases both the synthesis of the receptor subunits and the processing and assembly of the subunits to form functional receptors.

Regulation of expression of extracellular matrix proteins is likely to be particularly important in the female reproductive tract. Dissolution of connective tissue around a follicle is necessary for ovulation to occur, and increased levels of connective tissue are involved in formation of the corpus luteum and growth of the uterine myometrium during pregnancy (44). The identification of TGFβ by immunohistochemical techniques in the mouse ovary and uterus (5) suggest it may play a role in regulation of extracellular matrix production in these tissues.

Effects of TGFβ on Immune Cells

TGFβ is a potent immunosuppressive agent. In vitro, it inhibits proliferation of both T- and B-lymphocytes and suppresses the secretion of IgG and IgM by B-cells even in the presence of interleukin-2 (45–46). Evidence indicates that TGFβ can also suppress immune function in vivo. Patients with glioblastoma often have impaired cell-mediated immune responses, and their serum can inhibit proliferation of normal T-lymphocytes. Purification of this immune suppressive factor from a glioblastoma cell line showed it to be $TGF\beta_2$ (8). Because of these immunosuppressive effects, TGFβ could be involved in acceptance of the embryo by the mother during pregnancy. Immunoreactive $TGF\beta_1$ has been found in both pre- and post-implantation mouse embryos (28, 47) and the uterus of pregnant mice (47), while high levels of immunoreactive $TGF\beta_2$ are in the 18-days' murine placenta (48). Additionally, an immunosuppressive molecule in supernatents from murine decidua is related to $TGF\beta_2$ (49). The immunosuppressive activity of this molecule can be neutralized by antibodies to $TGF\beta_2$, but not to $TGF\beta_1$, and is recognized on Western blots by a $TGF\beta_2$ antibody. The presence of $TGF\beta_2$, as opposed to $TGF\beta_1$, as an immunosuppressive agent active during pregnancy is advantageous since $TGF\beta_2$ does not inhibit endothelial cell growth in vitro and so would not be expected to inhibit the neovascularization required for fetal nourishment.

PARACRINE AND AUTOCRINE ACTIONS OF TGFβ IN THE REPRODUCTIVE TRACT

Besides being a potential mediator of growth and differentiation for cells in the reproductive tract, work by a number of investigators in the past four years indicates that TGFβ also has specialized functions in gonadal cells concerned with the regulation of steroidogenesis. Although initial studies have focused on the effects of exogenous TGFβ on these cells, studies now indicate that gonadal cells themselves synthesize TGFβs. TGFβ seems to affect steroid synthesis, and certain steroids have been shown to affect production of TGFβ by these cells. Thus, the gonads can serve as a model that demonstrates both autocrine and paracrine actions of TGFβ and the interrelationship between TGFβ and steroid hormones. Studies on actions of TGFβ on cells of the reproductive tract are reviewed next in this regard.

Actions of TGFβ in the Testis

Figure 2 summarizes the autocrine and paracrine actions of TGFβ in the testis. Treatment of cultured anterior pituitary cells with $TGF\beta_1$ causes increased secretion of follicle-stimulating hormone (FSH) with no effect on lutenizing hormone (LH) secretion (50). FSH action on Sertoli cells seems to decrease production of TGFβ by these cells (51). Cultures of porcine (52) and rat (51) Sertoli cells secrete a factor that has TGFβ bioactivity and that competes with iodinated TGFβ for receptor binding, while mRNA prepared from Sertoli cells hybridizes to a $TGF\beta_1$ probe (52). TGFβ does not affect the growth of these cells, nor does it seem to affect their differentiation, as it does not alter expression of the transferrin receptor (52). Even though FSH treatment decreases the basal secretion of TGFβ to undetectable levels, treatment of the cells with 10^{-7} M estradiol, dexamethasone, or thyroxine

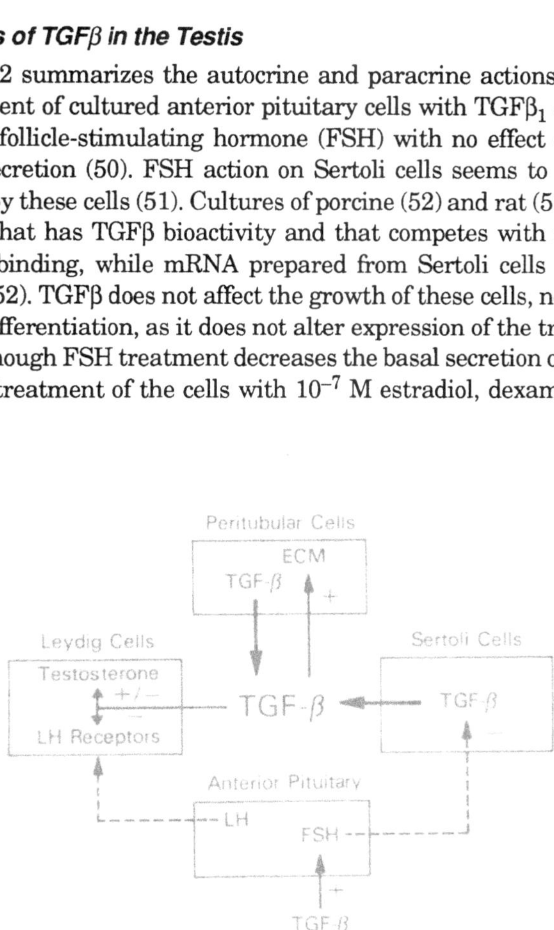

Fig. 2. Possible action of TGFβ in the testis. TGFβ is secreted by cultured Sertoli and peritubular cells (bold lines). Solid lines show the actions of TGFβ on given cells in vitro (+ implies increased expression; – is decreased expression; +/– indicates a biphasic response, where low concentrations of TGFβ increase expression and higher doses decrease expression). Dashed lines show effects of other agents on cells.

increases TGFβ levels in Sertoli cell-conditioned media by 4-, 6-, and 10-fold, respectively (51), indicating that steroids can regulate TGFβ production.

Cultures of rat peritubular cells (52) secrete TGFβs into the media, and the 2.4-kb TGFβ_1 mRNA is present in these cells. While TGFβ does not affect the growth of these cells, it does promote their migration and formation of colonies of cells in culture. TGFβ also increases secretion of a number of proteins by peritubular cells and may affect production of extracellular matrix protein by these cells. Thus, TGFβ produced by peritubular cells may have autocrine actions on these cells, while TGFβ secreted by Sertoli cells could act in a paracrine fashion on peritubular cells. Moreover, TGFβ produced by peritubular and Sertoli cells has been implicated in paracrine control of steroidogenesis in Leydig cells. As with other cell types in the testis, TGFβ does not affect proliferation of Leydig cells, which do possess TGFβ receptors (53). TGFβ reduces expression of human chorionic gonadotropin (hCG) receptors on porcine Leydig cells by almost 70% (53–54) and reduces cAMP response to hCG by almost 50% (54). It also affects steroid synthesis directly (55) since conversion of exogenous pregnenolone to testosterone is increased in cells treated with TGFβ, even though the steroidogenic response to hCG is depressed (54). The effect of TGFβ on testosterone production may be biphasic, with doses of TGFβ lower than 2 ng/ml stimulating testosterone production and higher doses inhibiting synthesis (53).

Actions of TGFβ in the Ovary and Uterus

In the ovary, cultures of both granulosa cells and thecal cells secrete TGFβ and TGFβ acts on both cell types, as shown in Figure 3. Whether the actions of TGFβ are of an autocrine or paracrine nature is not known. Again, TGFβ may be involved in increasing production of FSH by the anterior pituitary (50), and TGFβ seems to further synergize with FSH in granulosa cells, where it potentiates FSH-mediated increases in LH receptor levels, aromatase activity, and progesterone production. In cultured mouse (56) and rat (57) granulosa cells, treatment with 10 ng/mL TGFβ has no effect on aromatase activity, while treatment with FSH causes a 100-fold increase in estrogen accumulation. The combination of FSH plus TGFβ gives a 2- to 4-fold increase in aromatase activity as compared to FSH alone. TGFβ does not alter FSH binding to granulosa cells, but does increase forskolin-stimulated cAMP production, as well as cAMP-supported estrogen generation (56). Furthermore, TGFβ does not seem to have significant effects on granulosa cell number in the presence or absence of FSH (56, 58), but one study (59) indicates that TGFβ in the presence of FSH can increase [^{3}H]thymidine incorporation into cells. TGFβ does inhibit EGF-induced mitosis of granulosa cells (60). TGFβ at ng/mL concentrations also increases FSH-stimulated progesterone production by 2- to 3-fold (61–62). TGFβ alone has no effect on LH receptor numbers (61–62), but has a biphasic effect in the presence of FSH: At low concentrations of FSH (5 ng/mL) TGFβ augments the increase in LH receptors by 2.5-fold, but at optimal doses of FSH (50 ng/mL), TGFβ addition decreases the number of LH receptors by 50% (62). In these studies (61–62), TGFβ has little effect on FSH-induced cAMP production, in contrast to studies by Adashi and coworkers (56). Differences in specific experimental condi-

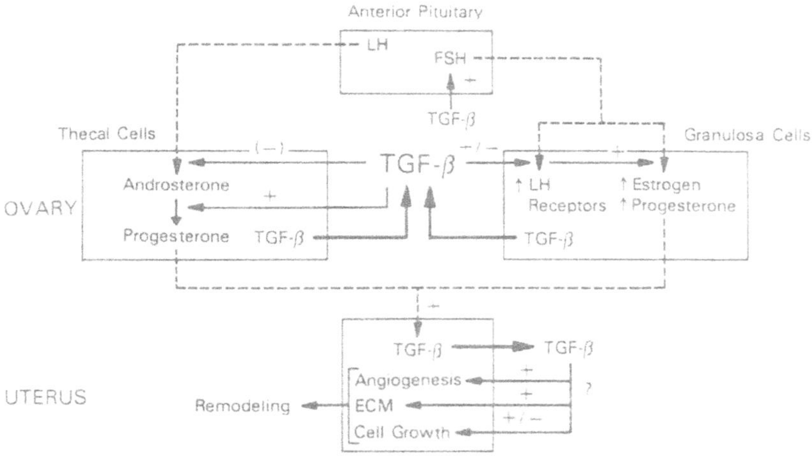

Fig. 3. Possible actions of TGFβ in the ovary and uterus. TGFβ is secreted by cultured granulosa and thecal cells and has been identified in the uterus in vivo (bold lines). Solid lines indicate TGFβ actions on cultured ovarian cells and postulated actions in the uterus (+ is a stimulatory effect by TGFβ; – is an inhibitory effect; +/– is a mixed effect). Dashed lines represent actions of other agents on cells.

tions may contribute to this discrepancy, or it may be due, in part, to use of systems where granulosa cell adenylate cyclase is already maximally stimulated by FSH.

Cultured rat granulosa cells secrete both active and latent TGFβ into their conditioned media, and there is an increased production of active TGFβ when cells are cultured on fibronectin-coated plates (63). Porcine granulosa cells produce significantly less TGFβ than rat granulosa cells. A preliminary study (64) also indicates that TGFβ is present in human follicular fluid, suggesting that granulosa cells in vivo may be a site of TGFβ synthesis. Additionally, rat thecal/interstitial cells (60, 63) and bovine thecal cells (60) in culture secrete TGFβ, and immunoreactive $TGF\beta_1$ has been localized to interstitial cells in the murine ovary (5). TGFβ does not affect the growth of rat thecal/interstitial cells (65) or basal levels of androsterone production (56, 65), but the effect on steroid synthesis in the presence of LH is not clear. While Adashi and coworkers (56) find no change in androsterone accumulation induced by 1-ng/mL hCG upon TGFβ addition, Magoffin and coworkers (65) show that concomitant treatment with LH (50 ng/mL) and TGFβ (10 ng/mL) causes a 65% reduction in androsterone accumulation compared to treatment with LH alone. They attribute the decrease in androsterone accumulation to a decrease in the activity of 17-alpha-hydroxylase/$C_{17\text{-}20}$ activity brought about by TGFβ. Interestingly, LH-induced progesterone production is increased 10-fold by TGFβ in this system, possibly due to a reported increase in production of cholesterol side chain-cleaving enzyme induced by TGFβ.

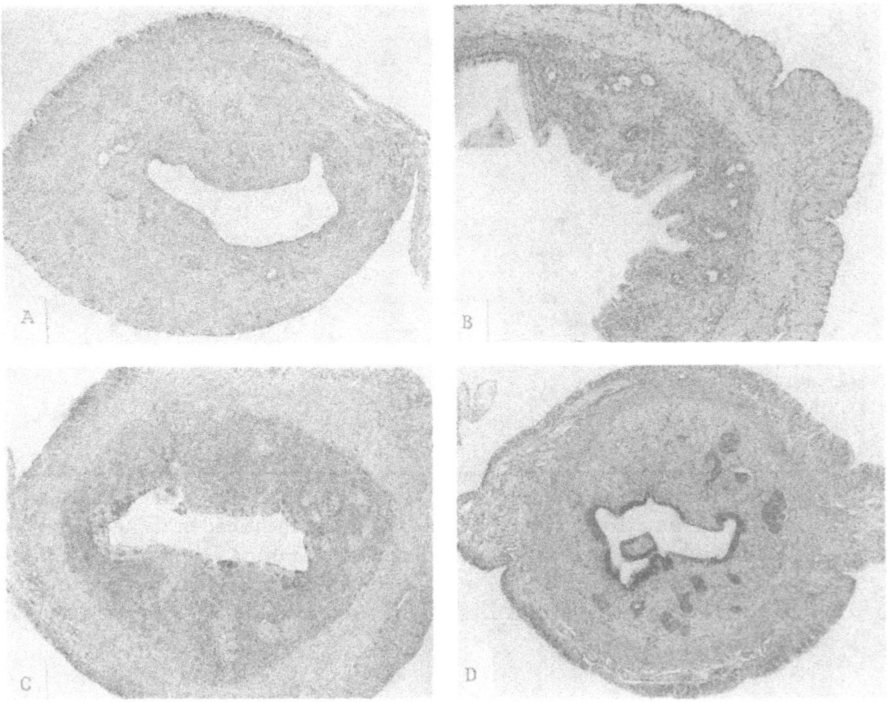

Fig. 4. Immunohistochemical localization of $TGF\beta_2$ in the uterus from an untreated rat (*A*) and rats treated with 17-β estradiol for 3 days (*B*), tamoxifen for 7 days (*C*), or progesterone for 7 days (*D*) (62×). Details of the treatment are given in the text. Tissues were immersion fixed in formalin and embedded in paraffin for sectioning. $TGF\beta_2$ was localized using an affinity-purified antibody raised to a peptide corresponding to amino acids 50–75 of the mature $TGF\beta_2$ sequence, as described in reference 48. Immune complexes were visualized by avidin-biotin peroxidase and diaminobenzidine.

In the ovary it appears that both granulosa cells and thecal/interstitial cells synthesize TGFβ, and while TGFβ does not have dramatic growth effects on these cells, it can alter steroid synthesis in both cell types. Studies are now demonstrating that steroids can also alter production of TGFβ. Tamoxifen induces $TGF\beta_2$ secretion by the human prostatic carcinoma cell line, PC-3[66]; antiestrogens increase secretion of active $TGF\beta_1$ from fetal fibroblasts (L. Wakefield and A. Colletta, personal communication); and estrogen induces the expression of $TGF\beta_1$ mRNA in human HOS TE85 osteosarcoma cells (67). Initial studies also indicate that TGFβ production in the uterus may be regulated by levels of steroid hormones (Fig. 3). Using immunohistochemical and in situ hybridization techniques, the preimplantation uterus of a pregnant mouse seems to synthesize TGFβ$_1$ in the glandular and luminal epithelium, while TGFβ$_1$ protein accumulates in the extracellular matrix of the stroma surrounding the epithelium (47). After implantation, the primary decidual zone appears to be synthesizing the TGFβ that accumulates in the matrix of the secondary decidual zone and decidua capsularis.

We have begun to look at the expression of TGFβs 1, 2, and 3 in the uterus in response to treatment with steroids. Ten days after oophorectomy, rats were given daily injections of 17-β estradiol (10 μg/kg), progesterone (10 mg/kg), or tamoxifen (500 μg/kg), and after 1, 3, and 7 days, uteri were removed and processed for immunohistochemical study. Controls show light staining for the TGFβs throughout the uterus (Fig. 4*A*). Estradiol (Fig. 4*B*) and tamoxifen (Fig. 4*C*) cause an induction of the TGFβs in endometrium, stroma, and myometrium, with the most pronounced induction in the stroma. The response was detected at 24 h, was greater at 3 days, and was maintained at 7 days. Tamoxifen at 3 days shows accentuated glandular staining as compared to estradiol. In contrast, progesterone (Fig. 4*D*) leads to a dramatic increase in TGFβ expression in the glandular epithelium over a similar time course. There is little increase in stromal staining. The expression patterns and induction of TGFβs 1, 2, and 3 are similar. In situ hybridization studies and Northern blot analyses are currently in progress.

Thus, the levels of TGFβ in the uterus may be regulated by estrogen and progesterone produced by the ovary (Fig. 3). Increased levels of TGFβ produced in response to steroid hormones may affect angiogenesis, extracellular matrix production, and cell growth and differentiation in the uterus. The significance of TGFβ induction in either glandular epithelium or stromal cells by different steroids is not understood. Because TGFβ can have profound effects on the growth and differentiation of both epithelial and mesenchymal cells types, a combination of autocrine and paracrine modes of action for TGFβ in the uterus are likely. Determination of the biological actions of TGFβ in uterine cells will require further in vitro studies.

SUMMARY

TGFβ appears to be an important component in the regulation of reproductive physiology. It can affect steroid synthesis and response in vitro, and the expression of TGFβ itself may be regulated by steroids. The ability of TGFβ to induce angiogenesis and production of extracellular matrix proteins may also be functionally important; however, these and other biological actions of TGFβ, many of which have been demonstrated in vitro, may be altered in vivo by other growth factors. In vivo work suggests that several TGFβ isoforms are expressed in reproductive organs; their differential actions there, as well as the regulation of their expression, remain to be investigated.

REFERENCES

1. Roberts AB, Anzano MA, Lamb LC, Smith JM, Sporn MB. New class of transforming growth factor potentiated by epidermal growth factor. Proc Natl Acad Sci USA 1981;78:5339-43.

2. Roberts AB, Sporn MB. The transforming growth factor-betas. In: Sporn MB, Roberts AB, eds. Handbook of experimental pharmacology. Heidelberg: Springer-Verlag, 1990;95:419-72.
3. Wakefield LM, Smith DM, Masui T, Harris CC, Sporn MB. Distribution and modulation of the cellular receptors for transforming growth factor-beta. J Cell Biol 1987; 105:965-75.
4. Heine UI, Flanders KC, Roberts AB, Munoz EF, Sporn MB. Role of transforming growth factor-β in the development of the mouse embryo. J Cell Biol 1987;105:2861-76.
5. Thompson NL, Flanders KC, Smith JM, Ellingsworth LR, Roberts AB, Sporn MB. Cell type specific expression of transforming growth factor-beta 1 in adult and neonatal mouse tissue. J Cell Biol 1989;108:661-9.
6. Derynck R, Jarrett JA, Chen EY, et al. Human transforming growth factor-beta cDNA sequence and expression in tumor cell lines. Nature 1985;316:4377-9.
7. Madisen L, Webb NR, Rose TM, et al. Transforming growth factor-β2: cDNA cloning and sequence analysis. DNA 1988;7:1-8.
8. de Martin R, Haendler B, Hofer-Warbinek R, et al. Complementary DNA for human glioblastoma-derived T cell suppressor factor, a novel member of the transforming growth factor-β gene family. EMBO J 1987;6:3673-7.
9. Jakowlew SB, Dillard PJ, Sporn MB, Roberts AB. Complementary deoxyribonucleic acid cloning of mRNA encoding transforming growth factor-β2 from chicken embryo chondrocytes. Growth Factors (in press).
10. Dijke P, Hanson P, Iwata KK, Pieler C, Foulkes JG. Identification of a new member of the transforming growth factor-β gene family. Proc Natl Acad Sci USA 1988;85: 4715-9.
11. Derynck R, Lindquist PB, Lee A, et al. A new type of transforming growth factor-β, TGF-β3. EMBO J 1988;7:3737-43.
12. Denhez F, Lafyatis R, Kondaiah P, Roberts AB, Sporn MB. Cloning by polymerase chain reaction of a new mouse TGF-β, mTGF-β3. Growth Factors (in press).
13. Jakowlew SB, Dillard PJ, Kondaiah P, Sporn MB, Roberts AB. Complementary deoxyribonucleic acid cloning of a novel transforming growth factor-β messenger ribonucleic acid from chick embryo chondrocytes. Mol Endocrinol 198;2:747-55.
14. Jakowlew SB, Dillard PJ, Sporn MB, Roberts AB. Complementary deoxyribonucleic acid cloning of an mRNA encoding transforming growth factor-beta 4 from chicken embryo chondrocytes. Mol Endocrinol 1988;2:1186-95.
15. Kondaiah P, Sands MJ, Smith JM, et al. Identification of a novel transforming growth factor β (TGF-β5) mRNA in *Xenopus laevis*. J Biol Chem 1990;265:1089-93.
16. Roberts AB, Kondaiah P, Rosa F, et al. Mesoderm induction in *Xenopus laevis* distinguishes between the various TGF-β isoforms. Growth Factors (submitted).
17. Roberts AB, Rosa F, Roche NS, et al. Isolation and characterization of TGF-β2 and TGF-β5 from medium conditioned by *Xenopus* XTC cells. Growth Factors (in press).
18. Jennings JC, Mohan S, Linkhart TA, Widstrom R, Baylink DJ. Comparison of the biological activities of TGF-β1 and TGF-β2: Differential activity in endothelial cells. J Cell Physiol 1988;137:167-72.
19. Glick AB, Flanders KC, Danielpour D, Yuspa SH, Sporn MB. Retinoic acid induces transforming growth factor-β2 in cultured keratinocytes and mouse epidermis. Cell Regulation 1989;1:87-97.

20. Mason AJ, Hayflick JS, Ling N, et al. Complementary DNA sequences of ovarian follicular fluid inhibin show precursor structure and homology with transforming growth factor-β. Nature 1985;318:659-63.
21. Ling N, Ying SY, Ueno N, et al. Pituitary FSH is released by a heterodimer of the β-subunits from the two forms of inhibin. Nature 1986;321:779-82.
22. Cate RL, Mattaliano RJ, Hession C, et al. Isolation of the bovine and human genes for mullerian inhibiting substance and expression of the human gene in animal cells. Cell 1986;45:685-98.
23. Padgett RW, St. Johnston RD, Gelbart WM. A transcript from a *Drosophila* pattern gene predicts a protein homologous to the transforming growth factor-beta family. Nature 1987;325:81-4.
24. Weeks DL, Melton DA. A maternal mRNA localized to the vegetal hemisphere in *Xenopus* eggs codes for a growth factor related to TGF-β. Cell 1987;51:861-7.
25. Lyons K, Graycar JL, Lee A, et al. Vgr-1, a mammalian gene related to *Xenopus* Vg-1, is a member of the transforming growth factor β gene superfamily. Proc Natl Acad Sci USA 1989;86:4554-8.
26. Wozney JM, Rosen V, Celeste AJ, et al. Novel regulators of bone formation: Molecular clones and activities. Science 1988;242:1528-34.
27. Kimelman D, Kirschner M. Synergistic induction of mesoderm by FGF and TGF-β and the identification of an mRNA coding for FGF in the early *Xenopus* embryo. Cell 1987;51:869-77.
28. Rappolee DA, Brenner CA, Schultz R, Mark D, Werb Z. Developmental expression of PDGF, TGF-α, and TGF-β genes in preimplantation mouse embryos. Science 1988;242:1823-5.
29. Lehnert SA, Akhurst RJ. Embryonic expression pattern of TGF-beta type 1 RNA suggests both paracrine and autocrine mechanisms of action. Development 1988; 104:263-73.
30. Robey PG, Young MF, Flanders, KC, et al. Osteoblasts synthesize and respond to TGF-beta in vitro. J Cell Biol 1987;105:457-63.
31. Seyedin SM, Thomas TC, Thompson AY, Rosen DM, Piez KA. Purification and characterization of two cartilage-inducing factors from bovine demineralized bone. Proc Natl Acad Sci USA 1985:82:2267-71.
32. Sporn MB, Roberts AB. Peptide growth factors are multifunctional. Nature 1988; 332:217-9.
33. Wakefield LM, Smith DL, Flanders KC, Sporn MB. Latent transforming growth factor-β from human platelets. J Biol Chem 1988;263:7646-54.
34. Rosen DM, Stempien SA, Thompson AY, Seyedin PR. Transforming growth factor-beta modulates the expression of osteoblast and chondroblast phenotypes in vitro. J Cell Physiol 1988;134:337-46.
35. Silberstein GB, Daniel CW. Reversible inhibition of mammary gland growth by transforming growth factor-β. Science 1987;237:291-3.
36. Gabrielson EW, Gerwin BI, Harris CC, Roberts AB, Sporn MB, Lechner JF. Stimulation of DNA synthesis in cultured primary human mesothelial cells by specific growth factor. FASEB J 1988;2:2717-21.
37. Takahashi T, Nelson K, Goods L, McLachlan JA. Transforming growth factor-β promotes the growth of mouse uterus and vagina, in vivo [Abstract]. J Cell Biochem 1989;13B:201.

38. Madri JA, Pratt BM, Tucker A. Phenotypic modulation of endothelial cells by transforming growth factor-β depends upon the composition and organization of the extracellular matrix. J Cell Biol 1988;106:1375-84.
39. Postlethwaite AE, Keski-Oja J, Moses HL, Kang AH. Stimulation of the chemotactic migration of human fibroblasts by transforming growth factor beta. J Exp Med 1987;165:251-6.
40. Ignotz RA, Massague J. Transforming growth factor-beta stimulates the expression of fibronectin and collagen and their incorporation into extracellular matrix. J Biol Chem 1986;261:4337-45.
41. Rossi P, Karsenty G, Roberts AB, Roche NS, Sporn MB, de Crombrugghe B. A nuclear factor 1 binding site mediates the transcriptional activation of a type I collagen promoter by transforming growth factor-β. Cell 1988;52:405-14.
42. Raghow R, Postlethwaite AE, Keski-Oja J, Moses HL, Kang AH. Transforming growth factor-β increases steady state levels of type I procollagen and fibronectin messenger RNAs posttranscriptionally in cultured human dermal fibroblasts. J Clin Invest 1987;79:1285-8.
43. Roberts CJ, Birkenmeier TM, McQuillan JJ, et al. Transforming growth factor-β stimulates the expression of fibronectin and of both subunits of the human fibronectin receptor by cultured human lung fibroblasts. J Biol Chem 1988;263:4586-92.
44. Junqueira LC, Carneiro J, Long JA, eds. Basic histology. Norwalk, CT: Appleton-Century-Crofts, 1986.
45. Kehrl JH, Roberts AB, Wakefield LM, Jakowlew SB, Sporn MB, Fauci AS. Transforming growth factor beta is an important immunomodulatory protein for human B-lymphocytes. J Immunol 1986;137:3855-60.
46. Kehrl JH, Wakefield LM, Roberts AB, et al. Production of transforming growth factor beta by human T lymphocytes and its potential role in the regulation of T cell growth. J Exp Med 1986;163:1037-50.
47. Tamada H, McMaster MT, Flanders KC, Andrews GK, Dey SK. Cell type-specific expression of TGF-β in the mouse uterus during the periimplantation period. Mol Endocrinol (submitted).
48. Flanders KC, Cissel DS, Mullen LT, Danielpour D, Sporn MB, Roberts AB. Antibodies to transforming growth factor-β2 peptides: Specific detection of TGF-β in immunoassays. Growth Factors (in press).
49. Clark DA, Flanders KC, Banwatt D, et al. Suppressor cells in murine pregnancy decidua produce an immunosuppressive molecule related to transforming growth factor β-2. J Immunol (in press).
50. Ying S-Y, Becker A, Baird A, et al. Type beta transforming growth factor (TGF-β) is a potent stimulator of the basal secretion of follicle stimulating hormone (FSH) in a pituitary monolayer system. Biochem Biophys Res Commun 1986;135:950-6.
51. Benahmed M, Cochet C, Kermidas M, Chauvin MA, Morera AM. Evidence for a FSH dependent secretion of a receptor reactive transforming growth factor-β-like material by immature Sertoli cells in primary culture. Biochem Biophys Res Commun 1988; 154:1222-31.
52. Skinner MK, Moses HL. Transforming growth factor-β gene expression and action in the seminiferous tubule: Peritubular cell-Sertoli cell interactions. Mol Endocrinol 1989;3:625-34.
53. Morera AM, Cochet C, Keramidas M, Chauvin MA, de Peretti E, Benahmed M. Direct

regulating effects of transforming growth factor β on the Leydig cell steroidogenesis in primary culture. J Steroid Biochem 1988;30:443-7.

54. Avallet O, Vigier M, Perrard-Sapori MH, Saez JM. Transforming growth factor β inhibits Leydig cell functions. Biochem Biophys Res Commun 1987;146:575-81.
55. Lin T, Blaisdell J, Haskell JF. Transforming growth factor-β inhibits Leydig cell steroidogenesis in primary culture. Biochem Biophys Res Commun 1987;146:387-94.
56. Adashi EY, Resnick CE, Hernandez ER, May JV, Purchio AF, Twardzik DR. Ovarian transforming growth factor-β (TGF-β): Cellular site(s) and mechanism(s) of action. Mol Cell Endocrinol 1989;61:247-56.
57. Ying S-Y, Becker A, Ling N, Ueno N, Guillemin R. Inhibin and beta type transforming growth factor (TGF-β) have opposite modulating effects on the follicle stimulating hormone (FSH)-induced aromatase activity of cultured rat granulosa cells. Biochem Biophys Res Commun 1986;136:969-75.
58. Feng P, Catt KJ, Knecht M. Transforming growth factor β regulates the inhibitory actions of epidermal growth factor during granulosa cell differentiation. J Biol Chem 1986;261:14167-70.
59. Dorrington J, Chuma AV, Bendell JJ. Transforming growth factor β and follicle-stimulating hormone promote rat granulosa cell proliferation. Endocrinology 1988;123:353-9.
60. Skinner MK, Keski-Oja J, Osteen KG, Moses HL. Ovarian thecal cells produce transforming growth factor-β which can regulate granulosa cell growth. Endocrinology 1987;121:786-92.
61. Knecht M, Feng P, Catt K. Bifunctional role of transforming growth factor-β during granulosa cell development. Endocrinology 1987;120:1243-9.
62. Dodson WC, Schomberg DW. The effect of transforming growth factor-β on follicle-stimulating hormone induced differentiation of cultured rat granulosa cells. Endocrinology 1987;120:512-6.
63. Kim I-C, Schomberg DW. The production of transforming growth factor-β activity by rat granulosa cell cultures. Endocrinology 1989;124:1345-51.
64. Ruegsegger Veit C, Assoian RK. Identification of transforming growth factor-beta in human ovarian follicular fluid [Abstract]. Endocrinology 1988;122 (suppl):1227.
65. Magoffin DA, Gancedo B, Erickson GF. Transforming growth factor-β promotes differentiation of ovarian thecal-interstitial cells but inhibits androgen production. Endocrinology 1989;125:1951-8.
66. Ikeda T, Lioubin MN, Marquardt H. Human transforming growth factor type β2; production by a prostatic adenocarcinoma cell line, purification, and initial characterization. Biochemistry 1987;26:4337-45.
67. Komm BS, Terpening CM, Benz DJ, et al. Estrogen binding, receptor mRNA, and biologic response in osteoblast-like osteosarcoma cells. Science 1988;241:81-4.

3

Transforming Growth Factor α: Expression and Biological Activities of the Secreted and Integral Membrane Forms

David C. Lee

Lineberger Cancer Research Center, Department of Microbiology and Immunology, University of North Carolina at Chapel Hill

Transforming growth factor α (TGFα) is a potent polypeptide mitogen that was first identified in the culture fluids of retroviral (1) and chemically transformed (2) cells. Originally referred to as "sarcoma growth factor," it was eventually renamed in light of its ability to promote a phenotypic transformation of normal fibroblasts in culture (3). The designation "α" distinguishes this molecule from TGFβ, a completely unrelated polypeptide that cooperates to enhance the TGFα-induced, anchorage-independent growth of normal fibroblasts in a soft agar assay (4).

Rat TGFα was purified from media conditioned by feline sarcoma virus (FeSV)-transformed cells, and its amino acid sequence was determined (5). It was found to be a 50-amino acid polypeptide with apparent structural homology to epidermal growth factor (EGF). This structural homology derives from the preservation of 6 cysteine residues and the resulting disulfide bonds, with some additional conservation, most notably in the third loop. This conservation of structure apparently allows TGFα to bind the mammalian EGF receptor with an affinity comparable to that of EGF (6–7). Moreover, binding similarly induces the receptor's intrinsic tyrosine kinase activity with subsequent down-regulation and loss of receptor from the cell surface (7–8).

TGFα and EGF are now known to be two members of a larger family of related proteins that also includes a growth factor encoded by vaccinia virus (VVGF) (9), as well as the newly discovered amphiregulin (10), a product of human breast cancer cells. These proteins share the invariant spacing of 6 cysteine residues, together with the ability to bind and activate the EGF receptor. In addition, EGF/TGFα-like sequences, again characterized by the conserved spacing of 6 cysteine residues, are found in a significant number of extracellular or transmembrane proteins with diverse function (11). This list includes cell recognition molecules encoded by homeotic loci in *Drosophila* and the nematode, the LDL receptor and proteases active in the blood-clotting cascade. The role, or roles, of these EGF-like units, which often occur as repeats (as they do in the EGF precursor), is presently unknown. Finally, in addition to a multiplicity of related growth factors, recent work from several laboratories suggests the existence of a family of EGF-receptor-related molecules that includes the neu/HER2 protein (12), as well as a newly-discovered HER3 (13). These different EGF-receptor-like proteins may discriminate both qualitatively and quantitatively in the binding of various members of the EGF family, underscoring the rich complexity of both the regulation and actions of these growth factor molecules.

The cloning of rat cDNAs revealed that TGFα is encoded by a 4.5-kb mRNA that includes a short 5' untranslated region of 150 nucleotides, followed by the coding sequence (see below), and, finally, a long 3' untranslated region of approximately 3900 nucleotides (14–15). This transcript, which is derived from 6 exons that span approximately 85 kb of DNA, is expressed at high levels in a variety of rodent and human transformed cells and tumors (16). In addition, its expression can be induced in several cell types by tumor promoters (17–19) and estrogen (20), as well as by TGFα (or EGF) itself through an autoinduction (18–19, 21) phenomenon. Expression of TGFα protein and/or mRNA have also been detected in a variety of normal adult tissues, including the bovine anterior pituitary (22), keratinocytes (21), activated macrophages (23), and portions of the nervous system (24). Finally, expression of TGFα mRNA has been observed in preimplantation blastocysts (25) and certain tissues of the 9- and 10-day-old mouse embryo (26).

EXPRESSION OF TGFα IN THE RODENT DECIDUA

Shortly after cloning the rat TGFα in 1985, we reported relatively high levels of TGFα mRNA associated with crudely dissected rat fetuses around days 8 and 9 of gestation (27). Subsequent in situ hybridization analysis revealed that the high levels of TGFα mRNA reported were actually contributed by contaminating decidua and not by the embryo proper (28). This was further confirmed by the results of Northern blot analyses in which we compared the expression of the 4.5-kb TGFα transcript in samples of poly(A^+) derived from decidua, uterus, and fetus at days 10, 11, and 14 of gestation. Expression was predominant in the decidua, with lower levels detected in the uterus. In contrast, TGFα transcripts were not detected in fetal samples. We did not examine earlier stages by Northern analysis since it was not possible to obtain sufficient quantities of cleanly dissected fetus.

To determine whether the lower levels of TGFα mRNA detected in the uterine samples were the result of decidual contamination, we compared expression in these tissues at earlier gestational ages when they can be more clearly dissociated. Specifically, we examined expression in the decidua, the remaining uterus at the site of implantation, and the intersegmental regions of the uterus between implantation sites at days 7, 8, 9, and 10 of gestation. At these ages we found significant expression of TGFα transcripts in the decidua, with little, if any, expression in the uterus.

Results from the aforementioned study indicated that the highest levels of expression occured at day 8 of gestation. This was confirmed by a more comprehensive study in which we compared the levels of TGFα mRNA in the decidua from days 7 through 11 of gestation. In addition, we also examined poly(A^+) RNAs from the nonpregnant uterus, the pregnant uterus at day 5 just prior to decidualization, and the uterus at days 14 and 15 when the decidua has largely resorbed and all that remains is the decidua basalis. The results of this analysis confirmed that the peak of expression in the decidua occurs around day 8 of gestation, at which point the levels of TGFα transcript are significantly higher than those seen in other peripheral tissues of the pregnant rat. Transcripts were not detected in the nonpregnant uterus (though in collaboration with John McLachlan's laboratory, we have shown that TGFα expression in the nonpregnant mouse uterus can be induced by estrogen treatment). An intriguing aspect to the decidual expression was the finding of an apparent gradient, with the highest levels occuring immediately adjacent to the embryo. The significance of this gradient is presently unknown.

The function of TGFα produced in the decidua is uncertain. Since both the decidua and uterus express EGF receptor (29), it is possible that the growth factor acts to stimulate mitosis through autocrine and paracrine mechanisms. In addition, since EGF (and presumably, therefore, TGFα) is known to induce prolactin in a rat pituitary cell line (30), and decidual cells produce a prolactin-like molecule (31), TGFα may be responsible for this specific induction of gene expression. Furthermore, TGFα is reported to influence cell migration (32); thus, its expression in the decidua (especially as a gradient) may help direct the influx of fetal cells. Finally, it is also possible that TGFα produced in the decidua is localized to the fetus through presently unknown transport mechanisms.

BIOLOGICAL PROPERTIES OF THE TRANSMEMBRANE TGFα PRECURSOR

Analysis of the human (33) and rat (14) TGFα cDNAs suggested that the mature growth factor is proteolytically cleaved from a glycosylated precursor protein of 159 or 160 amino acids. The prediction that this precursor (termed proTGFα) contains both a signal peptide and a second extremely hydrophobic domain with the characteristics of a transmembrane domain suggested that proTGFα is an integral membrane glycoprotein. The mature, 50-amino acid growth factor is apparently cleaved from the extracellular portion of this precursor through the action of an elastase-like enzyme that cleaves alanine-valine bonds at both termini. The salient features of this model have been confirmed biochemically in studies from several laboratories, including our own working in collaboration with Joan Massague, Larry Gentry, and colleagues (34–36).

The processing of TGFα from an integral membrane precursor may be a mechanism shared by certain other growth factors. For example, other members of the EGF family of growth factors are predicted (on the basis of nucleotide sequence) to be similarly cleaved from integral membrane precursors (37). In addition, experimental evidence confirms that colony-stimulating factor I (CSF-I) is cleaved from a transmembrane molecule (38). These observations have provoked considerable interest in the biological activities of these larger, membrane-spanning forms of growth factors. The existence of membrane-anchored growth factors might provide a mechanism whereby the actions of these potent molecules can—in the absence of the required proteases—be limited to adjacent cells. For example, such localized action might be important during development when inductive signals must be highly controlled. Alternatively, these molecules may have additional functions, perhaps as cell recognition or adhesion molecules. As a first step toward addressing these possibilities, we wished to determine whether membrane-anchored proTGFα can, in the absence of processing, productively interact with EGF receptor on adjacent cells.

To block proteolytic release of mature TGFα, we used a strategy of sequential mutagenesis (39). Since the specificity of the processing enzyme is unknown, and both alanine-valine cleavage sites are embedded in alanine/valine-rich sequences, we were concerned about the possibility of residual cleavage in the flanking regions. Accordingly, we constructed a series of mutants with increasing substitutions. Mutant 1-3 was constructed so as to eliminate both of the alanine-valine dipeptides that must be cleaved to release mature growth factor. Mutant 2-4 was constructed to contain additional substitutions that also eliminate residual valine/alanine and alanine/alanine dipeptides in the flanking sequences. Finally, to exclude the potential trypsin-like cleavage of a lysine-lysine bond between the mature growth factor sequence and the transmembrane domain, we created mutant 2-4T that in addition to the above substitutions, also contained substitutions at this dipeptide. However, we subsequently failed to find evidence of cleavage at this position even when cells that express the wild-type TGFα are incubated with concentrations of trypsin that rapidly release cells from the substrate. Thus, most of our detailed characterization has focused on mutant 2-4.

Wild-type and mutant cDNAs were cloned into expression vectors under the control of the heavy-metal-inducible, metallothionein-1 promoter. These vectors were transfected into baby hamster kidney (BHK) cells since we had previously shown them to process proTGFα to the 50-amino acid form (36), and they express essentially undetectable levels of EGF receptor. Selected clones were shown to be highly inducible for vector-derived TGFα transcripts, as shown by Northern blot analysis, and to express the wild-type and mutant proTGFα molecules by immunoprecipitation of ^{35}S-cysteine-labeled cells using an antibody directed against the cytoplasmic domain of the molecule. Two species of 18 and 21 kD were immunoprecipitated from cells expressing the wild-type cDNA; the 21-kD product was shown to result from O-linked glycosylation of the smaller species. In a pulse-chase experiment, the 21-kD product disappeared first, followed by the 18-kD band, and these were replaced by a new species of 16–17 kD. The latter eventually disappeared after a 2-h chase. With mutant 2-4, in contrast, an initial 18-kD molecule was rapidly converted to a diffuse band of 23 kD that was stable and not detectably processed even with long chase periods. The larger size of this mutant proTGFα was explained by the fact that it is more extensively glycosylated. This, presumably, is due to the substitution of serine and threonine residues for the wild-type sequence. Indirect immunofluorescence with a monoclonal antibody directed against mature growth factor sequence also confirmed expression and indicated that the mutations did not alter the surface localization of the precursor. Specific TGFα staining was observed around the peripheral membranes of induced mutant 2-4 cells and, to a lesser extent, those expressing wild-type proTGFα, but was not observed with parental BHK cells.

To verify that the amino acid substitutions blocked secretion of TGFα, media conditioned by induced wild-type and mutant clones were concentrated and tested for bioactivity and immunoreactivity using three sensitive methods: competition for EGF receptor binding, ability to stimulate the growth of normal fibroblasts in soft agar, and Western blot analysis. The results of these assays, which were entirely consistent, revealed that TGFα activity was readily detectable in culture supernatants from cells expressing wild-type proTGFα and, to a lesser extent, mutant 1-3, but not from cells expressing the mutant 2-4 construct. They substantiated our concern regarding residual cleavage in the flanking regions, and confirmed that the more extensively mutated proTGFα expressed in the 2-4 clones is resistant to processing.

Although larger secreted forms of TGFα are biologically active (40–41), the ability of intact proTGFα to bind and activate EGF receptor had not been previously established. As a first step in examining the biological activity of the precursor, we showed that solubilized wild-type proTGFα, immunoprecipitated from cells expressing the wild-type sequence using an antibody directed against the cytoplasmic domain, was able to bind to, and induce autophosphorylation of, solubilized EGF receptor from human A431 cells. We further showed that the intact mutant 2-4 precursor was similarly active. These results, then, allowed us to address the more important question of whether intact, membrane-anchored precursor on the surface of one cell can bind and activate EGF receptor on the surface of an adjacent cell. The experimental approach was to coincubate induced mutant 2-4 cells with serum-starved A431 cells for varying periods of time at 37°C. To probe for induced autophosphorylation of the A431 receptor, the cells were lysed, the lysates resolved by SDS-PAGE and transferred to nitrocellulose, and the blot probed with affinity-purified antiphophotyrosine antisera. Binding of the primary antibody was then visualized with ^{125}I-Protein A. Using this approach, we found a reproducible increase in the phosphotyrosine content of the A431 receptor as early as 15 min after the addition of induced, but not uninduced, 2-4 cells. This increase was not observed when A431 cells were incubated in the presence of mutant 2-4 cell wash supernatants, and in control experiments with [^{35}S]methionine-labeled 2-4 cells, we could not detect any apparent degradation of the precursor.

The aforementioned results indicate that membrane-anchored growth factor can bind to, and activate, EGF receptor on an adjacent cell surface. To determine whether this interaction leads to immediate downstream signal transduction, we examined a well-characterized early response to soluble growth factor; namely, the transient rise in free intracellular Ca^{++}. Induced 2-4 cells were washed and coincubated with serum-starved A431 cells that had been preloaded with the fluorescent calcium indicator FURA-2. We then monitored changes in FURA-2 fluorescence by digital fluorescence video microscopy in collaboration with Brian Herman of the University of North Carolina. We found that within 15–30 s after adding the induced 2-4 cells, there was a 5- to 7-fold increase in the A431 cell-free intracellular Ca^{++}. As is the case with soluble growth factor, this was a transient response, and free Ca^{++} levels had returned to baseline within 5 min. We observed a much smaller increase in free Ca^{++} when A431 cells were coincubated with uninduced 2-4 cells and no increase following coincubation with the parental BHK cells.

We have recently observed a similar 2-4-induced increase in free Ca^{++} in primary rat hepatocytes cultured in the presence of EGTA. This result is consistent with an IP_3-mediated release of Ca^{++} from intracellular stores rather than an influx of Ca^{++} from the medium. Finally, using a special computer program to monitor individual cells, we have further shown that while not all hepatocytes in a given field respond initially, additional cells show increased fluorescence when the overlying 2-4 cells are redistributed by gentle pipetting. This is consistent with the establishment of new contacts between growth factors and receptors on the surface of the respective cells.

The previous results demonstrate immediate downstream signal transduction following interaction of membrane-anchored proTGFα and EGF receptor in situ (39). Similar results have recently been reported by Derynck and colleagues (42). To examine for possible late actions, we assayed for transforming activity (43). We constructed a series of retroviral expression vectors that placed either wild-type or mutant TGFα cDNAs under the control of the 5' LTR of the murine leukemia virus and used these vectors to infect normal rat kidney (NRK) cells. Clones of cells harboring either the parental or TGFα expression vectors were then selected and characterized with respect to growth in soft agar and tumorigenicity in nude mice. For these studies, we also constructed a retroviral vector containing a mutant sequence not previously described. This proTGFα mutant, designated *4*, contains the same amino acid substitutions as the 2-4 construct, but only at the carboxy-terminal cleavage site. Thus, it is designed to prevent release of the mature growth factor from the cell surface, but allow the removal of additional NH_2-terminal sequence, including the glycosylation site. We chose to construct this additional mutant since our studies of precursor processing suggested that similar forms of proTGFα might accumulate on the cell surface.

Characterization of NRK clones infected with these various retroviral vectors revealed that whereas cells harboring the parental vector continued to grow in a normal contact-inhibited, anchorage-dependent manner, clones that expressed wild-type proTGFα formed foci in culture and large colonies in soft agar. Moreover, clones expressing wild-type proTGFα had formed tumors at 20 out of 32 injection sites by week 4, and these tumors persisted throughout the course of the experiment. Expression of the two mutant forms of proTGFα (4 and 2-4) also allowed growth of NRK cells in soft agar, but only in the presence of a higher-than-normal serum concentration (20% versus 10%). In the case of mutant 4, the resulting colonies were at least as large as those induced by wild-type proTGFα, whereas colonies induced by the more extensively mutated 2-4 molecule were not as large.

A similar distinction was observed with the tumorigenicity assay. Clones expressing mutant 4 formed tumors at exactly the same frequency as did those expressing wild-type precursor (20/32 injection sites at week 4). Mutant 2-4 cells, on the other hand, also formed tumors, but with a longer latency. Thus, at week 4 tumors had formed at 4 out of 32 injection sites, but this number had increased to 11 out of 32 sites by week 8. In contrast, uninfected NRK cells, or those infected with the parental retroviral vector, never formed tumors. It is interesting to note that the average size of tumors induced by the two mutant forms of proTGFα was somewhat larger than that produced by the wild-type molecule. We have established cell cultures from these various tumors and shown that these cultures are resistant to neomycin (neomycin resistance is encoded by the retroviral vector) and that they express the appropriately sized retroviral transcript, confirming that they indeed arose from the injected cell population. More importantly, we have established by Western blot analysis that whereas cells derived from wild-type proTGFα tumors secrete mature 6-kD TGFα into the medium, those derived from the mutant proTGFα tumors do not.

The above described results demonstrate that interaction between membrane-anchored proTGFα and EGF receptor leads to downstream signal transduction and that it can result in the stimulation of DNA synthesis as indicated by transformation (42). They, thus, confirm the biological activity of the integral-membrane form of this particular growth factor and may have implications regarding the activity of membrane-anchored forms of other growth factors. Finally, they also suggest that the accumulation of proTGFα on the surface of some transformed cells (41) may have pathological significance. Whether proTGFα also has physiological significance remains to be determined.

EXPRESSION OF TGFα AND PROTGFα IN TRANSGENIC MICE

In collaboration with Eric Sandgren, Richard Palmiter, and Ralph Brinster, we have recently developed lines of transgenic mice that overexpress TGFα (44). Our goals in developing such mice were to address two issues: First, does overexpression of TGFα influence the normal program of cell differentiation and growth that directs fetal and postnatal development, and if so, which tissues and cell types are affected? Second, given the demonstrated ability of TGFα to transform fibroblasts in vitro and its expression in carcinomas in vivo, would it act as a transforming protein in vivo, particularly in epithelial tissues from which most cancers arise?

To obtain significant expression, we found it necessary to utilize constructs that contained exon/intron. Thus, in one case, a rat TGFα cDNA was fused to the first exon of the human growth hormone gene such that the growth hormone reading frame would not be expressed. In the other case, a portion of the TGFα cDNA was replaced with a fragment of genomic DNA (15) containing exon 5 and portions of exons 4 and 6, thereby generating a minigene. In both cases, expression was driven from the zinc-inducible mouse metallothionein-I promoter. Each of these various constructs was microinjected into fertilized mouse eggs that were then transferred to pseudopregnant foster females according to established procedures. Positive founder mice were used to establish several lines of mice for each construct. Since similar results were obtained with the fusion and minigene constructs, their use in the following experiments will not necessarily be distinguished.

Northern blot analyses of total RNA demonstrated a high level of expression of the metallothionein-driven TGFα/hGH fusion construct in several adult tissues, including liver, pancreas, kidney, and regions of the intestine. Transgene mRNA was also detected at lower levels in the coagulation gland, stomach, and nonvirgin mammary gland. In all cases, the level of expression was markedly enhanced by administration of zinc sulfate in water, and in no case did we detect transcripts in tissues from control mice. Surprisingly, despite this high level of expression and the previous demonstration that the metallothionein promoter is active in development, overexpression did not discernably affect fetal development either in terms of the numbers of pups born or their tissue morphology.

In contrast, induced expression from birth resulted in the enlargement of several adult organs as reflected by increased wet weight relative to controls. Thus, although total body weight actually declined slightly, the liver, pancreas, coagulation gland, and portions of the gastrointestinal tract were enlarged up to 3-fold in the transgenic animals. This enlargement was, in all cases, accompanied by similar increases in DNA content consistent with an increase in cell number (hyperplasia) rather than an increase in cell mass (hypertrophy). A comparison of several lines suggested that in general, there was a correlation between growth enhancement and the level of expression in a given tissue. However, the coagulation gland displayed comparatively modest expression relative to its increase in size. Even more striking was the finding that expression in the kidney was among the highest, and yet this organ usually displayed no significant increase in size. That this lack of response in kidney was not due to a failure to produce or process the TGFα precursor was demonstrated by Western blot analysis using a monoclonal antibody directed against a portion of mature TGFα. The results of this analysis confirmed that extracts of induced transgenic liver, kidney, and pancreas, but not the corresponding control tissues, contained the expected 6-kD immunoreactive protein that comigrated on SDS-PAGE with a synthetic 50-amino acid rat TGFα.

Although TGFα can enhance the proliferation of both epithelial and mesenchymal cells in culture, most of the enlarged organs in the transgenic mice displayed primarily epithelial hyperplasia. For example, in liver there was an increase in the amount of histologically normal parenchyma, and the colon displayed mucosal hyperplasia, with deeper, more cellular glands. Similarly, the duodenum contained larger, thicker villi that contained more cells. In each of these organs, the cells appeared relatively normal, and overall tissue architecture was preserved.

In certain tissues, however, the response to chronic TGFα overexpression was not limited to ordered epithelial hyperplasia but, in fact, included examples of metaplasia, dysplasia, and even neoplasia. For example, compared to the normal pancreas, the transgenic organ exhibited a dramatic increase in the number of fibroblasts, as well as extracellular matrix deposition, resulting in widely separated acini. Most acini and individual acinar cells appeared normal, although scattered throughout the affected pancreas were focal clusters of ductlike structures. These structures, which have been previously observed in rodents treated with certain pancreatic carcinogens and in human pancreatic tumors, may represent preneoplastic lesions (45). Described as pseudoductular acinar metaplasia (or tubular ductal complexes), they appear to result from either the de- or redifferentiation of entire acini. This interpretation is supported by the finding that individual cells within these ductlike structures frequently stain positive for amylase. Thus, in the pancreas, chronic elevated expression of this growth factor induced proliferation of both mesenchyme and epithelium and also appeared to focally redirect epithelial differentiation.

To determine whether the marked pancreatic phenotype is reproduced by local TGFα expression, we generated transgenic mice bearing a rat elastase enhancer/promoter fused to the TGFα cDNA/hGH gene construct. In these mice, highly specific transgene expression is targeted to pancreatic acinar cells. These mice displayed all of the pancreatic changes previously described. Indeed, the pancreas of one 3-month old mouse was enlarged 10-fold relative to controls. That the pancreases of these mice exhibited similar severe fibrosis and multifocal pseudoductular acinar metaplasia argues that these effects are largely the result of autocrine and paracrine actions. Conversely, that phenotypes in other tissues (see below) were not observed in mice expressing these elastase-driven constructs argues that endocrine actions are not significant in this context.

As described above, the transmembrane precursor of TGFα has demonstrated biological activity in fibroblasts in culture. To determine its influence on pathology in the pancreas, the noncleavable 2-4 mutant proTGFα construct, driven by the elastase promoter, was also overexpressed in transgenic mice. The pancreas of these animals exhibited some, but not all, of the changes elicited by the wild-type proTGFα transgene. Specifically, the 2-4 mutant proTGFα induced multifocal pseudoductular acinar metaplasia in the absence of fibroplasia. The dissociation of these effects indicates that secretion of TGFα is necessary for paracrine growth stimulation of fibroblasts in this system. Furthermore, these results confirm that proTGFα, as a membrane-anchored ligand, can modulate differentiation and proliferation of contiguous epithelial cells in vivo.

Changes in the transgenic coagulation gland, which elaborates components of the postcoital plug, were even more striking. The coagulation glands of the transgenic mice were massively enlarged relative to normal controls. Histologically, unlike the normal coagulation gland, which is a well-ordered structure containing fingerlike projections of 2-cell-thick secretory epithelium, the transgenic gland displayed cystic mucosal hyperplasia with multiple large papillae and basal cystic spaces. The cells comprising the secretory epithelium were often pleiomorphic and dysplastic, with occasional mitotic figures. In advanced lesions, nests of fibroblasts were visible within a thickened epithelial stroma, and in some areas focal dysplastic changes were compatible with carcinoma in situ.

Finally, that TGFα can act as a frank oncoprotein was clearly demonstrated by the response in breast tissue. A transgenic mouse that had raised 6 litters developed a secretory mammary adenocarcinoma at 14 months of age. This tumor contained dysplastic to anaplastic alveolar cells with bizarre, pleomorphic nuclei and nucleoli, and scant to extremely vacuolated cytoplasm. Ductal and alveolar lumens were generally found to contain secretions despite the fact that this mouse had ceased lactating several months prior to tissue collection. Similar though less-extreme lesions were observed in focal areas of mammary glands in mice that had raised 1–3 litters. Mammary tissue in these mice displayed numerous focal collections of cells resembling the hyperplastic alveolar nodules that are occasionally observed in aged female mice and are believed to be possible precursors to mammary tumors. However, the cells in the nodules of the transgenic animals were extremely dysplastic, consistent with the presence of carcinoma. Finally, a single transgenic mouse developed hepatocellular carcinoma at 9 months of age. Although this result has not yet been reproduced, tumors of this type have not been previously seen in the nontransgenic Brinster mouse colony, suggesting that expression of TGFα contributed to its development.

The aforementioned results demonstrate that TGFα functions as both a potent epithelial cell mitogen and an oncoprotein in vivo. In addition, they are consistent with earlier suggestions that TGFα plays a significant role in mammary carcinogenesis (46). Finally, they also raise interesting questions regarding the basis of the differing tissue response. For example, why do fibroblasts in the pancreas respond to transgene expression of TGFα, whereas those in the liver and kidney do not? Does this have a trivial explanation (e.g., differences in the circulation in these tissues), or is there a fundamental difference in the responsiveness of fibroblasts in various tissues? Also, what are the underlying differences that account for the fact that overexpression of TGFα yields ordered epithelial hyperplasia in some tissues, and yet significantly lower levels produce carcinoma in the breast. Finally, why does the kidney, which displays one of the highest levels of expression of TGFα and contains EGF receptor (47), not demonstrate a phenotype? Answers to these questions are fundamental to an understanding of the homeostatic mechanisms that regulate growth, as well as the ways in which growth factors impinge on these mechanisms.

REFERENCES

1. DeLarco JE, Todaro GJ. Growth factors from murine sarcoma virus-transformed cells. Proc Natl Acad Sci USA 1978;75:4001-5.
2. Liu C, Tsao MS, Grisham JW. Transforming growth factors produced by normal and neoplastically transformed rat liver epithelial cells in culture. Cancer Res 1988; 48:850-5.
3. Roberts AB, Lamb LC, Newton DL, et al. Transforming growth factors: Isolation of polypeptides from virally and chemically transformed cells by acid/ethanol extraction. Proc Natl Acad Sci USA 1980;77:3494-8.
4. Roberts AB, Anzano MA, Lamb LC, et al. New class of transforming growth factors potentiated by epidermal growth factor: Isolation from non-neoplastic tissues. Proc Natl Acad Sci USA 1981;78:5339-43.

5. Marquardt H, Rose TM, Webb NR, et al. Rat transforming growth factor type I: Structure and relation to epidermal growth factor. Science 1984;223:1079-82.
6. Todaro GJ, Fryling C, DeLarco JE. Transforming growth factors produced by certain human tumor cells: Polypeptides that interact with epidermal growth factor receptors. Proc Natl Acad Sci USA 1980;77:5258-62.
7. Massague J. Epidermal growth factor-like transforming growth factor, II. Interaction with epidermal growth factor receptors in human placenta membranes and A431 cells. J Biol Chem 1983;258:13614-20.
8. Pike LJ, Marquardt H, Todaro GJ, et al. Transforming growth factor and epidermal growth factor stimulate the phosphorylation of a synthetic, tyrosine-containing peptide in a similar manner. J Biol Chem 1982;257:14628-31.
9. Stroobant P, Rice AP, Gullick WJ, et al. Purification and characterization of vaccinia virus growth factor. Cell 1985;42:383-93.
10. Shoyab M, Plowman GD, McDonald VL, et al. Structure and function of human amphiregulin: A member of the epidermal growth factor family. Science 1989;243: 1074-6.
11. Bender W. Homeotic gene products as growth factors. Cell 1985;43:559-60.
12. Bargmann CI, Hung M-C, Weinberg RA. The *neu* oncogene encodes an epidermal growth factor receptor-related protein. Nature 1986;319:226-30.
13. Kraus MH, Issing W, Miki T, et al. Isolation and characterization of ERBB3, a third member of the ERBB/epidermal growth factor receptor family: Evidence for overexpression in a subset of human mammary tumors. Proc Natl Acad Sci USA 1989; 86:9193-7.
14. Lee DC, Rose TM, Webb NR, et al. Cloning and sequence analysis of a cDNA for rat transforming growth factor-α. Nature 1985;313:489-91.
15. Blasband AJ, Rogers KT, Chen X, et al. Characterization of the rat transforming growth factor-α gene and identification of promoter sequences. Mol Cell Biol 1990 (in press).
16. Derynck R, Goeddel DV, Ullrich A, et al. Synthesis of messenger RNAs for transforming growth factors α and β and the epidermal growth factor receptor by human tumors. Cancer Res 1987;47:707-12.
17. Pittelkow MR, Lindquist PB, Abraham RT, et al. Induction of transforming growth factor-α expression in human keratinocytes by phorbol esters. J Biol Chem 1989;264: 5164 -71.
18. Raymond VW, Lee DC, Grisham JW, et al. Regulation of transforming growth factor α messenger RNA expression in a chemically transformed rat hepatic epithelial cell line by phorbol ester and hormones. Cancer Res 1989;49:3608-12.
19. Mueller SG, Kobrin MS, Paterson AJ, et al. Transforming growth factor-α expression in the anterior pituitary gland: Regulation by epidermal growth factor and phorbol ester in dispersed cells. Mol Endocrinol 1989;3:976-83.
20. Liu SC, Sanfilippo B, Perroteau I, et al. Expression of transforming growth factor α (TGFα) in differentiated rat mammary tumors: Estrogen induction of TGFα production. Mol Endocrinol 1987;1:683-92.
21. Coffey RJ, Derynck R, Wilcox JN, et al. Production and auto-induction of transforming growth factor-α in human keratinocytes. Nature 1987;328:817-20.
22. Kobrin MS, Samsoondar J, Kudlow JE. α-Transforming growth factor secreted by untransformed bovine anterior pituitary cells in culture, II. J Biol Chem 1986;261: 14414-9.

23. Madtes DK, Raines EW, Sakariassen KS, et al. Induction of transforming growth factor-α in activated human alveolar macrophages. Cell 1988;53:285-93.
24. Wilcox JN, Derynck R. Localization of cells synthesizing transforming growth factor-alpha mRNA in the mouse brain. J Neurosci 1988;8:1901-4.
25. Rappolee DA, Brenner CA, Schultz R. Developmental expression of PDGF, TGF-α, and TGF-β genes in preimplantation mouse embryos. Science 1988;241:1823-5.
26. Wilcox JN, Derynck R. Developmental expression of transforming growth factors alpha and beta in mouse fetus. Mol Cell Biol 1988;8:3415-22.
27. Lee DC, Rochford R, Todaro GJ, et al. Developmental expression of rat transforming growth factor-alpha mRNA. Mol Cell Biol 1985;5:3644-6.
28. Han VKM, Hunter ES, Pratt RM, et al. Expression of rat transforming growth factor alpha mRNA during development occurs predominantly in the maternal decidua. Mol Cell Biol 1987;7:2335-43.
29. Chegini N, Rao, CV. Epidermal growth factor binding to human amnion, chorion, decidua and placenta from mid- and term pregnancy: Quantitative light microscopic autoradiographic studies. J Clin Endocrinol Metab 1985;61:529-35.
30. Murdoch GH, Potter E, Nikolaisen AK, et al. Epidermal growth factor rapidly stimulates prolactin gene transcription. Nature 1982;300:192-4.
31. Golander A, Hurley T, Barrett J, et al. Synthesis of human prolactin by decidua in vitro. J Endocrinol 1979;82:263-8.
32. Barrandon Y, Green H. Cell migration is essential for sustained growth of keratinocyte colonies: The roles of transforming growth factor-α and epidermal growth factor. Cell 1987;50:1131-7.
33. Derynck R, Roberts AB, Winkler ME, et al. Human transforming growth factor-α: Precursor structure and expression in *E. coli*. Cell 1984;38:287-97.
34. Bringman TS, Lindquist PB, Derynck R. Different transforming growth factor-α species are derived from a glycosylated and palmitoylated transmembrane precursor. Cell 1987;48:429-40.
35. Teixido J, Gilmore R, Lee DC, et al. Integral membrane glycoprotein properties of the prohormone pro-transforming growth factor-α. Nature 1987;326:883-5.
36. Gentry LE, Twardzik DR, Lim GJ, et al. Expression and characterization of transforming growth factor α precursor protein in transfected mammalian cells. Mol Cell Biol 1987;7:1585-91.
37. Gray A, Dull TJ, Ullrich A. Nucleotide sequence of epidermal growth factor cDNA predicts a 128,000-molecular weight protein precursor. Nature 1983;303:722-5.
38. Rettenmier CW, Roussel MF, Ashmun RA, et al. Synthesis of membrane-bound colony stimulating factor 1 (CSF-1) and downmodulation of CSF-1 receptors in NIH 3T3 cells transformed by cotransfection of the human CSF-1 and *c-fms* (CSF-1) receptor genes. Mol Cell Biol 1987;7:2378-87.
39. Wong ST, Winchell LF, McCune BK, et al. The TGFα precursor expressed on the cell surface binds to the EGF receptor on adjacent cells, leading to signal transduction. Cell 1989;56:495-506.
40. Ignotz RA, Kelly B, Davis RJ, et al. Biologically active precursor for transforming growth factor type α released by retrovirally transformed cells. Proc Natl Acad Sci USA 1986;83:6307-11.
41. Luetteke NC, Michalopoulos GK, Teixido J, et al. Characterization of high molecular weight transforming growth factor α produced by rat hepatocellular carcinoma cells. Biochemistry 1988;27:6487-94.

42. Brachmann R, Lindquist PB, Nagashima M, et al. Transmembrane TGF-α precursors activate EGF/TGF-α receptors. Cell 1989;56:691-700.
43. Blasband AJ, Gilligan DM, Winchell LF, et al. Expression of the TGFα integral membrane precursor induces transformation of NRK cells (submitted).
44. Sandgren EP, Luetteke NC, Palmiter RD, et al. Overexpression of TGFα in transgenic mice: Induction of epithelial hyperplasia, pancreatic metaplasia, and carcinoma of the breast. Cell (in press).
45. Longnecker DS. Lesions induced in rodent pancreas by azaserine and other pancreatic carcinogens. Environ Health Perspec 1984;56:245-51.
46. Salomon DS, Zweibel JA, Bano M, et al. Presence of transforming growth factors in human breast cancer cells. Cancer Res 1984;44:4069-77.
47. Mydlo JH, Michaeli J, Cordon-Cardo C. Expression of transforming growth factor α and epidermal growth factor receptor messenger RNA in neoplastic and nonneoplastic human kidney tissue. Cancer Res 1989;49:3407-11.

PART

II

GROWTH FACTORS AND GONADAL FUNCTION

4

Growth Factor Regulation of Testicular Function

Brian P. Mullaney and Michael K. Skinner

Department of Pharmacology, Vanderbilt University School of Medicine, Nashville, Tennessee

The evolution of multicellular organisms required communication between different cells to coordinate tissue function. These cell-cell interactions have become vital for the regulation of cellular function, growth, and differentiation.

CELL-CELL INTERACTIONS IN THE TESTIS

The testis provides a useful model system to study cell-cell interactions due to the presence of a variety of cell types and the local production of regulatory factors. Testicular cell-cell interactions between Leydig, peritubular, Sertoli, and germ cells are important in the regulation of testis function and the process of spermatogenesis (1). Interactions between these different cell types may be categorized into environmental, nutritional, and regulatory interactions. Environmental interactions are mediated by such components as extracellular matrix and cell adhesion molecules. For example, Sertoli cells provide the proper microenvironment and cytoarchitectural support for the developing germinal cells. Peritubular-myoid cells contribute to the exterior wall of the seminiferous tubule and are separated from Sertoli cells by an extracellular matrix. This extracellular matrix and tight junctions between Sertoli cells form the blood-testis barrier. Therefore, Sertoli cells have nutritional interactions with germinal cells through the production of transport proteins necessary for the delivery of essential metabolites across the blood-testis barrier. Regulatory type cell-cell interactions are also present in the testis and are mediated by paracrine factors via receptor-mediated signal transduction events. Growth factor regulation of cell growth and differentiation are considered regulatory type interactions. This chapter briefly reviews growth factor-mediated cell-cell interactions in the testis.

TESTICULAR GROWTH AND DEVELOPMENT

Growth regulation is necessary for the development of the testis and maintenance of spermatogenesis. In the prepubertal testis Sertoli cells divide and grow to form the seminiferous tubule. The growth of these cells arrests at early puberty, and these cells become terminally differentiated. Peritubular-myoid cells appear in late fetal development and further proliferate. Similarly, Leydig cells also appear in late fetal development, but then degenerate and may further proliferate with the coincident initiation of spermatogenesis. Germinal cell development begins shortly after birth when gonocytes mitotically divide, forming spermatogonia. Developing spermatogonia traverse the blood-testis barrier and mature. Postpubertal growth control of somatic and germinal cell types is also necessary. At the onset of puberty, germinal cell meiosis begins, and the developing spermatozoa become the most abundant cell type in the seminiferous tubule. Proliferation of somatic and germ cells may require growth factors; however, other signals must be present to terminate growth and initiate differentiation. For example, growth inhibitory factors must be responsible for halting Sertoli cell growth and stimulating differentiation. Therefore, both positive and negative growth regulation appear to be necessary for the development of testis function.

GROWTH FACTORS IN THE TESTIS

A complex variety of growth factors appears to be required for growth regulation in the testis (2). Seminiferous growth factor (SGF) was the first mitogenic factor identified in the tubule (3–4). Although this protein has not been fully characterized, initial biochemical studies suggest it is not a previously identified growth factor, but this remains to be thoroughly investigated. The presence of the blood-testis barrier may require local production of essential factors for germinal cell division due to the exclusion of agents normally found in serum. One such factor, insulin-like growth factor I (IGF-I), has been identified in whole testis (5). Immunological evidence suggests that Sertoli cells produce IGF-I (6), while receptors are present on Leydig, Sertoli, and germinal cells (7). IGF-I can stimulate cell function, as indicated by increased Leydig cell steroidogenesis and Sertoli cell transferrin production (8–10). Another general growth factor, basic fibroblast growth factor (bFGF), may locally regulate gonadal function (11–12). Studies indicate bFGF is mitogenic for immature porcine Sertoli cells (13–14). A neurotrophic factor, beta-nerve growth factor (β-NGF), is also expressed in the testis (15). NGF mRNA has been detected in developing germ cells, while NGF receptor message may be present on Sertoli cells (16). Testosterone down regulates NGF receptor mRNA levels in vivo (16). These observations suggest a potential germ cell-Sertoli cell interaction. Surprisingly, immunological growth peptides, namely, an interleukin-like factor (IL-I), have also been detected in the testis (17). IL-I may regulate growth or immune suppression in the seminiferous tubule. More recently, transforming growth factors alpha and beta (TGFα and TGFβ), growth regulators with antagonistic actons, have been shown to be produced locally in the testis (Fig. 1).

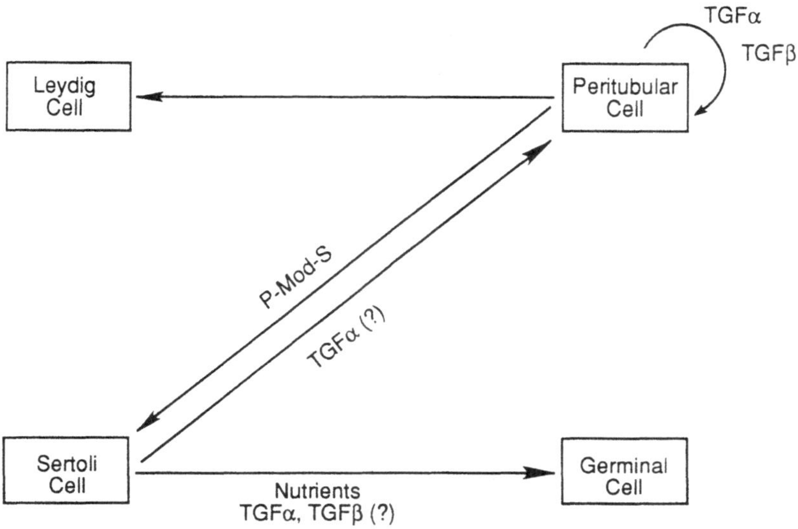

Fig. 1. Potential transforming growth factor-mediated cell-cell interactions in the testis.

Epidermal growth factor (EGF) has been implicated in the maintenance of spermatogenesis (18). Sialoadenectomized mice exhibited a 50% reduction of mature sperm, while EGF replacement returned spermatogenesis to normal levels. However, low circulating concentrations of EGF appear to imply the local production of an EGF-like factor. Local production of an EGF-like substance in the testis was initially supported by a report that Sertoli cells secrete a factor that blocks EGF from binding to its receptor (19). A candidate for mediating these effects is TGFα. TGFα is a peptide that has high homology with EGF, acts at the EGF receptor, and can mimic the actions of EGF (20). Therefore, the potential local production of TGFα as the EGF-like substance in the testis was examined. TGFα expression and protein production were found in midpubertal Sertoli and peritubular cells, but not in a crude mixed population of germ cells (21). Analysis of the potential sites of action of TGFα in the testis was investigated by localization of the EGF receptor. Scatchard analysis revealed high-affinity (100-pM) EGF receptor-binding sites on peritubular cells, but no EGF receptors were detected on midpubertal Sertoli cells (21). In contrast, another report presents immunological evidence that Sertoli cells may contain EGF receptors (22). Experimental limitations to be considered include the sensitivity of binding analysis, antibody crossreactivity, or possible expression of a nonfunctional truncated form of the receptor. The possibility that Sertoli cells contain EGF receptor is currently being examined with molecular probes for the receptor. The literature also suggests that EGF can alter Sertoli cell functions, such as lactate, inhibin, and estrogen production (23–24). Due to potential interactions between peritubular and Sertoli cells, some observed effects may be mediated through the peritubular cell contaminant of Sertoli cell preparations. For example, EGF can stimulate transferrin production in Sertoli-peritubular cocultures, but not in pure Sertoli cell cultures (21). Therefore, analysis of the actions of EGF/TGFα on Sertoli cell function requires further investigation.

To define further the growth role of TGFα, developmental studies were initiated on prepubertal, midpubertal, and mature isolated cells of the rat testis. Initial studies revealed that TGFα was expressed by both peritubular and Sertoli cells. TGFα can stimulate peritubular cell growth at all stages of development, as indicated by [^{3}H]thymidine incorporation into DNA and increases in cell number. However, neither adult nor immature Sertoli cells respond to TGFα, suggesting that another mitogen is responsible for immature Sertoli growth. Thus, peritubular cell proliferation may be jointly controlled by both peritubular cells and Sertoli cells. Observations have also demonstrated that Leydig cells are responsive to EGF and contain the EGF receptor. Therefore, EGF-mediated seminiferous tubule-Leydig cell interaction is a potential interaction that remains to be investigated. The role of TGFα in Sertoli cell-germ cell interactions is also unclear at present. If developing spermatogonia contain EGF receptors, however, Sertoli cell production of TGFα would provide an appropriate mechanism for initiating spermatogonial growth.

Growth inhibition also appears to be important in maintaining testis function. Terminal differentiation of Sertoli cells requires inhibition of Sertoli cell growth. In addition, the tightly regulated growth of germinal cells may also involve negative growth regulation. The presence of growth inhibitors may be necessary in the control of testis function. Therefore, the potential action of the growth inhibitor TGFβ in the testis was investigated. In comparison with TGFα, TGFβ is a multifunctional regulatory molecule. TGFβ generally inhibits EGF/TGFα-induced cell proliferation. TGFβ can also promote cellular differentiation, stimulate extracellular matrix production, and induce chemotaxis (25). Initial studies suggested that TGFβ-like proteins are produced by Sertoli cells, and their secretion may be modulated by gonadotropins (26). Northern analysis indicates that peritubular and Sertoli cells express TGFβ genes (27). The growth inhibitory action of TGFβ was examined in developmental studies. TGFβ can inhibit TGFα-stimulated peritubular cell growth in all stages of development, while having no effect on immature Sertoli cell growth. Local production of TGFβ may be a mechanism to limit TGFα-induced proliferation (27). The potential role of TGFβ to control the proliferation of spermatogonia in the adult and perhaps prevent prepubertal spermatogenesis remains to be investigated. The role of TGFβ as a differentiation factor is also being examined. Leydig cell steroidogenesis is decreased by TGFβ (28), while TGFβ has no effect on adult Sertoli cell function, including transferrin production (27). However, TGFβ may be important in peritubular cell differentiation. TGFβ can stimulate the production of several high molecular weight proteins, possibly matrix components, by peritubular cells. In vitro, TGFβ induces migration and colony formation of peritubular and Sertoli-peritubular cell cocultures (27). Therefore, morphogenesis and structural formation of the seminiferous tubule may be dependent on this factor. TGFβ-stimulated chemotaxis may also be a mechanism to recruit nondifferentiated peritubular cells to the exterior of the tubule during development. These observations imply that TGFβ may play an important role in cell-cell interactions in the testis (Fig. 1).

CELLULAR DIFFERENTIATION VERSUS GROWTH

While control of growth is necessary for development of testis function, control of differentiation is vital for development of specialized cellular function. A developing hypothesis is that the control of growth and differentiation are inversely linked. Thus, growth factors may act to shift the cell from a differentiated state to a less-differentiated growth state. For example, EGF and FGF decrease Leydig cell steroidogenesis, and EGF down-regulates hCG receptors (29–30). Therefore, growth factors appear to shift the cell away from its androgen-producing differentiated state, resulting in cell proliferation. Promotion of cellular differentiation will likely involve other nonmitogenic differentiation factors. As an example, a potential testicular differentiation factor in the seminiferous tubule, PModS, is produced by peritubular cells and modulates Sertoli cell differentiation (31). PModS enhances the majority of Sertoli cell functions, such as transferrin and inhibin production (32). PModS can stimulate Sertoli cell function to a greater extent than other known hormones, including FSH. This stimulation is due in part to a unique signal transduction mechanism involving cGMP (33). The production of PModS is regulated by androgens, suggesting a potential indirect mechanism for androgen action in the testis. This paracrine factor is postulated to be essential for the maintenance and control of normal testis function. Therefore, the postulate is made that growth factors may influence growth and development of the testis, while local production of differentiation factors may be important in maintaining testis function.

The growth and development of the testis involves differential growth of a number of cell types, including mesenchymal, epithelial, and germinal cells. In order to coordinate the temporal growth of these cells, a number of factors may be involved. Presently, a variety of growth factors have been identified in the testis. These factors include TGFα, TGFβ, SGF, IGF-I, FGF, NGF and IL-I. Local production and action of these factors may regulate the complex process of cell growth and differentiation. A controlled balance of growth and differentiation factors will likely be important in the development and maintenance of testicular function.

REFERENCES

1. Skinner MK. Cell-cell interactions in the testis. Ann NY Acad Sci 1987;513:158-71.
2. Bellve AR, Zheng W. Growth factors as autocrine and paracrine modulators of male gonadal functions. J Reprod Fertil 1989;85:771-93.
3. Feig LA, Bellve AR, Erikson NH, Klagsbrun M. Sertoli cells contain a mitogenic peptide. PNAS 1980;8:4774-8.
4. Bellve AR, Fieg LA. Cellular proliferation in the mammalian testis: biology of the seminiferous growth factor. Recent Prog Horm Res 1984;40:531-61.
5. Casella SJ, Smith EP, VanWyk JJ, et al. Isolation of rat testis cDNAs encoding an

insulin-like growth factor I precursor. DNA 1988;6:325-30.
6. Chaterlain PG, Naville D, Saez JM. Somatomedin-C/insulin-like growth factor-I-like material secreted by Sertoli cells in vitro: Characterization and regulation. Biochem Biophys Res Comm 1987;146:1009-17.
7. Vannelli BG, Barni T, Orlando C, Natali A, Serio M, Balboni GC. Insulin-like growth factor-I (IGF-I) and IGF-I receptor in human testis: An immunohistochemical study. Fertil Steril 1988;49:666-9.
8. Benahmed M, Morera AH, Cauvin MC, DePeretti E. Somatomedic C/insulin-like growth factor I as a possible intratesticular regulator of Leydig cell activity. Mol Cell Endocrinol 1987;50:69-77.
9. Skinner MK, Griswold MD. Multiplication stimulating activity (MSA) can substitute for insulin to stimulate the secretion of testicular transferrin by cultured Sertoli cells. Cell Biol Int Rep 1983;7:441.
10. Borland K, Mita M, Oppenheimer CL, Blinderman LA, Massague J, Hall PF, Czech MP. The actions of insulin-like growth factors I and II on cultured Sertoli cells. Endocrinology 1984;114:240.
11. Ueno N, Baird A, Esch F, Ling N, Guiellemin R. Isolation and partial characterization of a basic fibroblast growth factor from bovine testis. Mol Cell Endocrinol 1985;49: 189-94.
12. Gospodarowicz D, Ferrara N. Fibroblast growth factor and the control of pituitary and gonad development and function. J Steroid Biochem 1989;32:183-91.
13. Jaillard C, Chatelain PG, Saez JM. In vitro regulation of pig Sertoli cell growth and function: Effects of fibroblast growth factor and somatomedin C. Biol Reprod 1987: 665-74.
14. Smith EP, Hall SH, Monaco L, French FS, Wilson EM, Conti M. A rat Sertoli cell factor similar to bFGF increases c-fos messenger ribonucleic acid in culture Sertoli cells. Mol Endocrinol 1989;3:954-61.
15. Ayer-LeLievre CA, Olson L, Ebendal T, Hallbook F, Persson HT. Nerve growth factor mRNA and protein in the testis and epididymis of mouse and rat. PNAS 1988;85: 2628-32.
16. Persson H, Ayer-Le Lievre CA, Soder O, Villar MJ, Metsis M, Olson L, Ritzen M, Hokfelt T. Expression of beta-nerve growth factor receptor mRNA in Sertoli cells downregulated by testosterone. Science 1990;247:704-7.
17. Kahn SA, Soder O, Seyd V, Gustafsson K, Lindh M, Ritzen EM. The rat produces large amounts of an interleukin-I-like factor. Int J Androl 1987;10:495-503.
18. Tsutsumi O, Kurachi H, Oka T. A physiological role of epidermal growth factor in male reproductive function. Science 1986;233:975-7.
19. Holmes SD, Spotts G, Smith RG. Rat Sertoli cells secrete a growth factor that blocks epidermal growth factor (EGF) binding to its receptor. J Biol Chem 1986;9:4076-80.
20. Derynck R. Transforming growth factor-alpha. Cell 1988;54:593-5.
21. Skinner MK, Takacs K, Coffey RJ Jr. TGF-alpha expression in the seminiferous tubule. Endocrinology 1989;124:845-54.
22. Suarez-Quian CA, Dai M, Onida M, Kriss RM, Dym M. Epidermal growth factor receptor localization in the rat and monkey testes. Biol Reprod 1989;41:921-32.
23. Morris PL, Vale WW, Cappel S, Bardin CW. Inhibin production by primary Sertoli cell-enriched cultures: Regulation by follicle stimulating hormones, androgens, and epidermal growth factor. Endocrinology 1988;122:717-25.
24. Mallea LE, Machado AJ, Navaroli F, Rommerts FFG. Epidermal growth factor

stimulates lactate production and inhibits aromatization in cultured Sertoli cells from immature rats. Int J Androl 1986;9:201-8.

25. Roberts AB, Sporn MB. Transforming growth factor beta. Adv Cancer Res 1988; 51:107-45.
26. Benahmed M, Cochet C, Keramidas M, Chauvin MA, Morera AM. Evidence for a FSH dependent secretion of a receptor reactive transforming growth factor beta-like material by immature Sertoli cells in primary culture. Biochem Biophys Res Comm 1988;153:1222-31.
27. Skinner MK, Moses HL. Transforming growth factor beta gene expression and action in the seminiferous tubule. Mol Endocrinol 1989;3:625-34.
28. Lin T, Blaisdell J, Haskell JF. Transforming growth factor beta inhibits Leydig steroidogenesis in primary cultures. Endocrinology 1987;146:387-94.
29. Welsh TH Jr, Hseuh AJW. Mechanism of inhibitory action of epidermal growth factor on testicular androgen biosynthesis in vitro. Endocrinology 182;110:1498-1506.
30. Fauser BCJM, Baird A, Hseuh AJW. Fibroblast growth factor inhibits luteinizing hormone-stimulated androgen production by cultured rat testicular cells. Endocrinology 1988;123:2935-41.
31. Skinner MK, Fetterolf PM, Anthony CT. Purification of a paracrine factor, PModS, produced by testicular peritubular cells that modulates Sertoli cell function. J Biol Chem 1987;263:2884-90.
32. Skinner MK, McLachlan RI, Bremner WJ. Stimulation of Sertoli inhibin secretion by the testicular paracrine factor PModS. Mol Cell Endocrinol 1989;66:239-49.
33. Norton JN, Skinner MK. Regulation of Sertoli cell function and differentiation through the actions of a testicular paracrine factor, PModS. Endocrinology 1989;124: 2711-9.

5

Tumor Necrosis Factor α: Localization and Actions Within the Preovulatory Follicle

P. F. Terranova, K. F. Roby, M. Sancho-Tello, J. Weed, and R. Lyles

Departments of Physiology and Gynecology and Obstetrics, University of Kansas Medical Center, Kansas City, and Reproductive Resource Center of Greater Kansas City, Overland Park, Kansas

In 1985 we reported that the hamster ovary contained mast cells (1). This discovery led us to investigate the effects of histamine and serotonin on follicular and luteal function because these amines are the primary secretory products of mast cells. We extended our findings to the cow (2) and the human (3–4). Another major secretory product of mast cells is heparin, and it was evident from a previous study that heparin could synergize with fibroblast growth factor in promoting endothelial cell growth (5). Thus, we began our search in the ovary for growth factors that bind heparin. During the course of our studies, Gospodarowicz reported that the bovine corpus luteum contained FGF (6). During our purification scheme, using various concentrations of ammonium sulphate, we found a fraction that inhibited the growth of endothelial cells in vitro and reported this at the 1987 SSR meeting at Cornell. Subsequently, we published a full description of our findings (7–8). It was at this point that we asked the question: What ovarian factor might retard the growth of endothelial cells in vitro? In searching the literature we found a publication by Sato, et al., 1986, that described morphological changes in the endothelial cells treated with tumor necrosis factor alpha (TFNα) (9). Thus, with the assistance of Genentech, Inc., we secured monoclonal and polyclonal antibodies to human recombinant TNFα and embarked on our studies elucidating the presence of TNF in the ovary of the rat, human, and cow. The following discussion of the pertinent literature on TNF in the ovary describes its presence in specific ovarian compartments and its effects on ovarian function in vitro.

LOCALIZATION

Extraction

In our initial studies (7–8), we found an inhibitor of endothelial cell thymidine incorporation in vitro that was partially purified from cow ovaries using ammonium sulfate (AS) precipitation. Supernatant fluid from the 100,000-g pellet of freshly homogenized ovaries was subjected to stepwise AS precipitation. Precipitates were collected sequentially at 40%, 60%, 80%, and 95% AS saturation, and then each was dissolved, dialyzed (M_r 8000 cutoff), and examined in tissue culture for effects on cellular thymidine incorporation by cow pulmonary artery endothelial cells (CPAE) and mouse fibroblasts (L929 and 3T3). The 80% AS precipitate (ppt) inhibited the in vitro uptake of [^{3}H]thymidine by CPAE and L929 cells (data not shown for L929), but not 3T3 cells (Fig. 1). Heparin-Sepharose (HS) chromatography of the 80% AS ppt revealed that the inhibitory activity on CPAE and L929 cells did not bind to HS; the inhibitory fraction was found in the HS column breakthrough (80% BT). The 80% BT fraction reduced CPAE [^{3}H]thymidine uptake as determined by autoradiography and increased cellular uptake of trypan blue. Serial fractions from Sephacryl S-200 exclusion chromatography of the 80% BT contained CPAE inhibitory activity in the M_r range 30,000–50,000 (Fig. 2). The inhibitory activity on endothelial cells and L929 fibroblasts, the stimulation of [^{3}H]thymidine incorporation in 3T3 cells, and the nonreduced molecular weight range of that fraction are similar to those previously reported for (TNFα) (9–13).

The molecular weight range (30,000–50,000) of the 80% BT thymidine-incorporation inhibitory activity as determined by Sephacryl S-200 filtration is also similar to TNFα (14). TNFα has a molecular weight of 17,000 under reducing conditions (15–16) and 45,000 under nonreducing conditions (16–18). Fisch and Gifford (14) have shown that native, nonreduced TNFα elutes from Sephacryl S-200 at a molecular weight of about 48,000.

The extraction procedures utilized were a modification of those previously used for the extraction of endothelial cell growth regulators (6, 19, 20). TNFα has also been extracted from rabbit serum (21–22) and *E. coli* cell lysates (23) with AS precipitation. A similarity therefore exists between methods for TNF extraction and methods used for the extraction of the 80% inhibitory activity. Since TNFα has also been localized immunocytochemically to the follicular and luteal compartments of the cow ovary (24), it is possible that the 80% inhibitory activity is TNF or a TNF-like fraction. Western blot analysis and neutralization of bioactivity to TNFα are required, but sufficient quantities of the highly specific antisera are at present unavailable.

Immunocytochemistry

In the cow, TNF was observed in the corpus luteum (CL) and antral and atretic follicles (24). TNF in the CL was observed in the thecal cords radiating into the center of the CL from the periphery. A few lightly staining TNF-positive cells were also distributed throughout the CL in relatively low numbers compared to thecal cords. The granulosal cells of antral follicles contained TNF that was observed

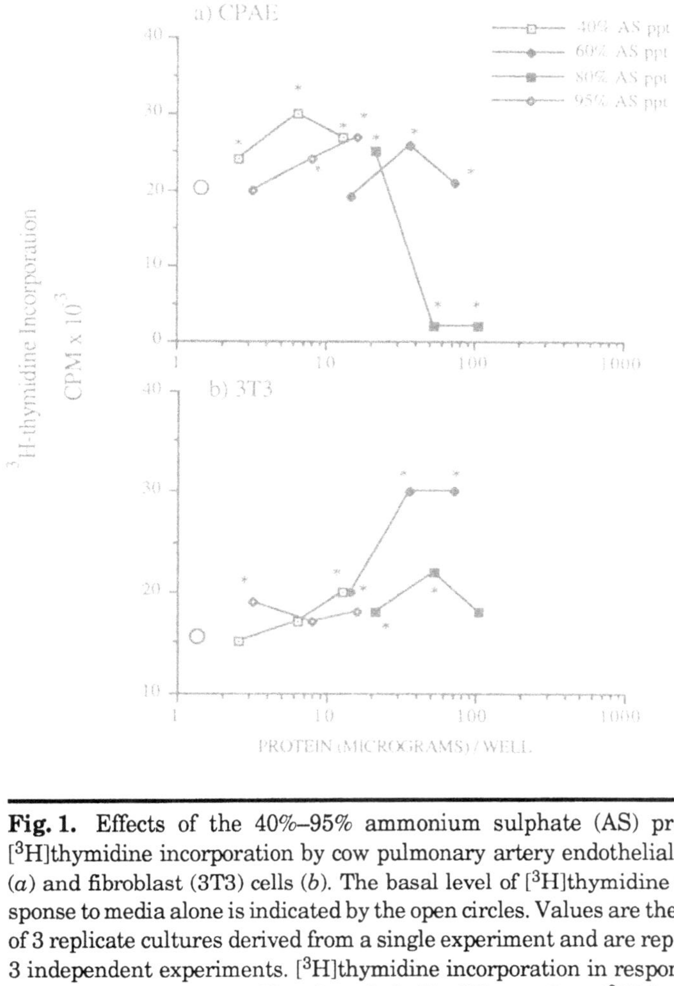

Fig. 1. Effects of the 40%–95% ammonium sulphate (AS) precipitates on [^{3}H]thymidine incorporation by cow pulmonary artery endothelial (CPAE) cells (*a*) and fibroblast (3T3) cells (*b*). The basal level of [^{3}H]thymidine uptake in response to media alone is indicated by the open circles. Values are the mean ± SEM of 3 replicate cultures derived from a single experiment and are representative of 3 independent experiments. [^{3}H]thymidine incorporation in response to various AS precipitates was considered statistically different from [^{3}H]thymidine incorporation in response to 0% FCS (control) when the P value was ≤0.05 as determined by Student's *t*-test. AS ppt vs. 0% FCS: *P < 0.05. (From Roby KF, Terranova PF, Partial purification of an endothelial cell growth regulator from the bovine ovary, J Reprod Fertil 1990;89:231–42.)

predominantly within the layers of granulosal cells lining the antral cavity and appeared to be in the follicular fluid surrounding the granulosal cells. TNF was observed in atretic follicles throughout the granulosal layer and in the fluid surrounding the granulosal cells. Sections incubated with preimmune rabbit serum showed no immunoreaction in any region of the ovary. A polyclonal antibody to human TNF neutralized with 100-fold excess TNF resulted in a reduction in the

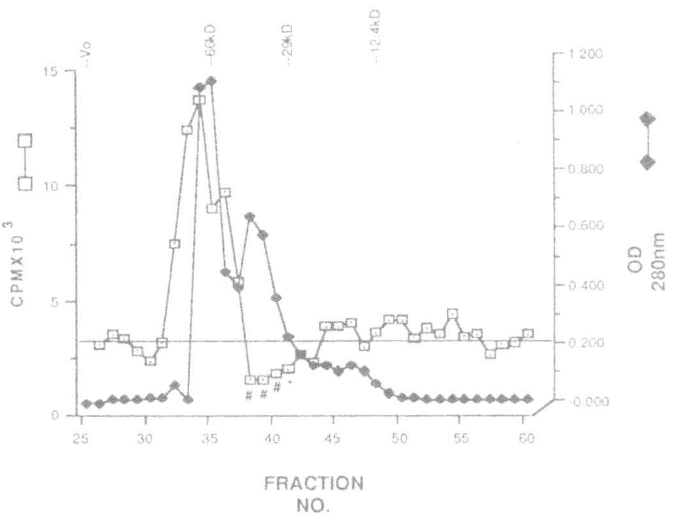

Fig. 2. The effect of Sephacryl S-200 chromatography of the 80% BT on CPAE cell thymidine incorporation. The 80% BT was applied to a Sephacryl S-200 column. The horizontal line indicates levels of [^{3}H]thymidine incorporation in response to media alone. An inhibitory region of activity corresponds to fractions 39–42, a molecular weight range of 30,000–50,000. The absorbance of each fraction at 280 nm is indicated (). Values are the mean ± SEM of 3 replicate cultures derived from a single experiment and are representative of 3 independent experiments. An inhibitory region was considered when at least two consecutive fractions exhibited a statistically significant reduction in [^{3}H]thymidine incorporation compared to the control of media alone. *$P < 0.05$ effect of treatment compared with control. (From Roby KF, Terranova PF, Partial purification of an endothelial cell growth regulator from the bovine ovary, J Reprod Fertil 1990; 89:231–42.)

intensity of immunostaining. In the presence of 1000-fold excess TNF, no immunostaining was apparent.

In the rat, TNF was localized in CL, throughout the granulosal layer in atretic follicles, and only in the more antral layers of granulosal cells in small and large preovulatory follicles (24). TNF was not apparent in preantral follicles. TNF in the CL was localized to the more diffuse cells in the central core. Qualitatively, the most intense TNF occurred in the granulosal layer of atretic follicles and in the granulosal cells lining the antral cavity, where TNF appeared to be in follicular fluid surrounding the granulosal cells.

In the human ovary (25), TNF was localized to the follicular and luteal compartments. Healthy antral follicles contained TNF in the antral layer of granulosa cells. TNF appeared to be secreted by the antral granulosa cells since it was located in the fluid surrounding these cells. TNF was apparent throughout the entire granulosa of atretic follicles and also appeared to be present in the follicular fluid surrounding the antral and pyknotic granulosa cells. The zona pellucida of atretic follicles contained TNF. TNF was apparent in the large lutein-like cells and in paraluteal cells of the CL and was located throughout the cytoplasm in the small paraluteal cells.

Rabbit. A study by Bagavandoss, et al. (26) has shown that rabbit luteal cell-conditioned media from days 5, 17, and 19 contained very low or undetectable TNFα activity. However, lipopolysaccharide (endotoxin) stimulated a marked increase in TNF activity of day 17–19 lutea cells. No TNF activity was detectable from unstimulated and lipopolysaccharide-stimulated nonluteal ovarian tissue and uterus. The activity of TNFα in the culture medium was neutralized by an antibody to rabbit TNFα.

EFFECTS OF TNF

Rat Follicles

In vitro TNF increased progesterone production by proestrous rat follicles compared to controls (27). Stepwise increases in progesterone production were observed with doses of TNF from 30 pM to 300 pM (Fig. 3). The highest dose of TNF (3000 pM) reduced progesterone production significantly compared to 300-pM TNF, but progesterone production was still higher than controls. Androstenedione production by ovarian follicles was reduced by 30-pM TNF and stimulated by 3000-pM TNF in vitro compared to controls. Estradiol production by ovarian follicles was unaffected by TNF in vitro during the 24 h of incubation.

A time course in response to 300-pM TNF revealed that progesterone production was low until 12 h; a significant increase was observed at 24 h. Neither androstenedione nor estradiol production by the follicles was altered by 300-pM TNF during the 24 h of incubation. LH (160 pM) increased progesterone and androstenedione at 6 h and both remained elevated through 24 h; estradiol was not affected by LH (data not shown).

TNF (30 pM) preabsorbed with 1000-fold excess of monoclonal antibody to TNF prevented the increase in progesterone production by the follicles compared to follicles incubated with 30-pM TNF. Follicles incubated with either no serum (controls), mouse serum, or 30-pM TNF absorbed with monoclonal antibody produced similar amounts of progesterone in vitro.

TNF significantly increased progesterone production by preovulatory rat follicles but not androstenedione or estradiol in the atmosphere of 5% CO_2 and air (28). Further analysis of the conditions of 5% CO_2 and air indicated that pregnenolone, 20α-dihydroprogesterone, and 17α-hydroxyprogesterone were also increased in response to TNF when compared to follicles cultured without TNF. TNF also increased production of progesterone when follicles were cultured in 5% CO_2 and 95% O_2. In contrast to follicles cultured with 5% CO_2 and air, a significant increase in the production of androstenedione and estradiol in response to TNF was observed by follicles cultured with 5% CO_2 and 95% O_2 (Fig. 4). Basal estradiol production, but not androstenedione production, was significantly increased by

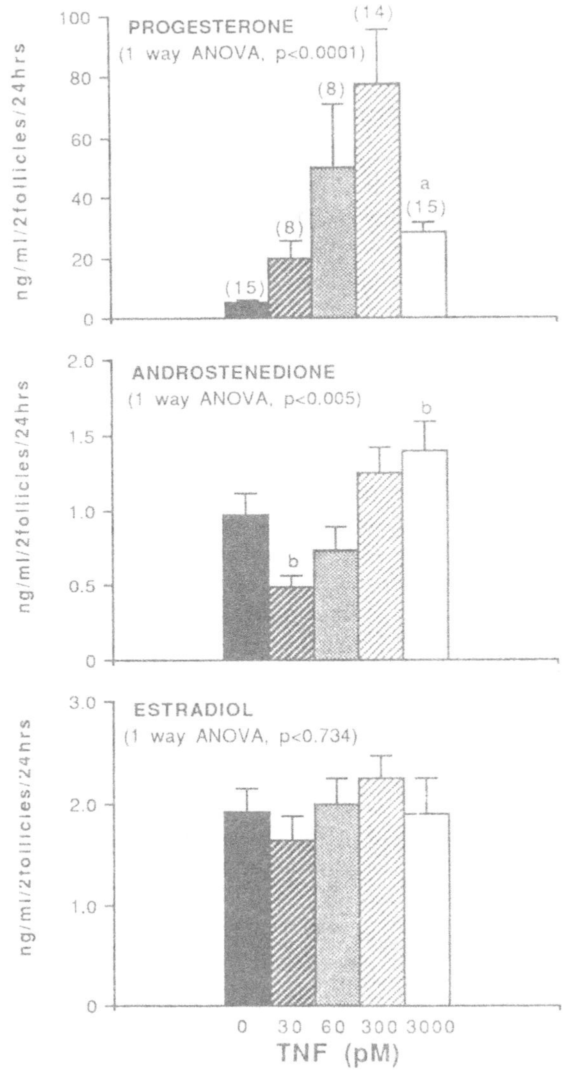

Fig. 3. Effects of 30- to 3000-pM TNF on rat follicular progesterone, androstenedione, and estradiol production in vitro. Two preovulatory follicles were incubated in 1-mL M199 with TNF for 24 h. (Numbers above bars are the number of replicates of each experiment; letters indicate levels of significance as follows: a = P < 0.05 vs. O-TNF, P < 0.001 vs. 300-pM TNF; b = P < 0.05 vs. O-TNF.) (From Roby KF, Terranova PF, Tumor necrosis factor alpha alters follicular steroidogenesis in vitro, Endocrinology 1990;123:2952–4.)

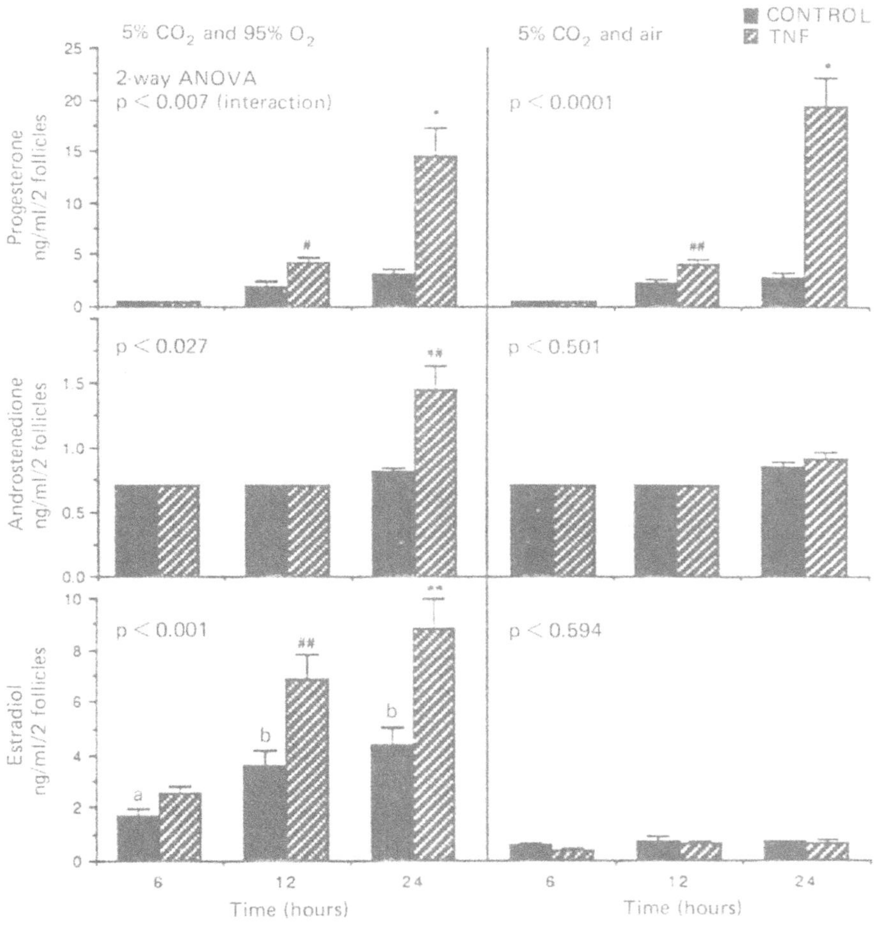

Fig. 4. Effect of TNF on follicular progesterone, androstenedione, and estradiol production at 6, 12, and 24 h of culture under the conditions of 5% CO_2 and 95% O_2 or 5% CO_2 and air. Steroids are expressed as ng/mL/2 follicles. The P values of the interaction between time and treatment were determined by 2-way ANOVA. Specific comparisons were performed using Student's *t*-test. (##, P < 0.025; #, P < 0.01; **, P < 0.005; *, P < 0.001, TNF vs. control. a: P < 0.025; b: P < 0.005, 5% CO_2 and 95% O_2 vs. 5% CO_2 and air for estradiol controls.) (From Roby KF, Terranova PF, Effects of tumor necrosis factor alpha in vitro on steroidogenesis of healthy and atretic follicles of the rat: theca as a target. Endocrinology 1990;126: 2711.)

follicles cultured in 5% CO_2 and 95% O_2 compared to those cultured in 5% CO_2 and air.

Morphology of the Preovulatory Rat Follicles: Effect of Culture Conditions

The effects of high and low oxygen in culture on the histology of the follicles was of interest to provide insight into the differences in in vitro steroidogenesis (28). Follicles cultured in 5% CO_2 and air exhibited extensive atresia. Follicular antra contained pyknotic granulosa cells. The cumulus was often expanded, and a degenerate oocyte containing condensed nuclear chromatin was observed near the center of the antrum. After 24 h of culture in 5% CO_2 and 95% O_2, follicles were healthy. The granulosal layer was intact, and only a few pyknotic granulosa cells were observed. The oocyte was at the periphery of the follicle and surrounded by a compact cumulus and oocyte with a distinct nuclear membrane. TNF did not alter the histology of the follicle in high or low O_2 during this 24 h incubation period.

Effect of TNF and Exogenous Steroids on Follicular Steroidogenesis

To ascertain if TNF was altering a specific enzymatic step in steroidogenesis, follicles were cultured in 5% CO_2 and air in the presence of exogenous steroid with or without TNF (28).

Addition of 25-hydroxycholesterol (0, 300, or 3000 ng/mL) did not alter pregnenolone production by the follicles in the absence of TNF (28). Addition of TNF in the presence of the increasing amounts of 25-hydroxycholesterol resulted in a dose-dependent increase in pregnenolone production. The increases were not accounted for by the effect of TNF on basal pregnenolone production in the absence of cholesterol.

Increasing concentrations of pregnenolone, progesterone, 17α-hydroxyprogesterone, or androstenedione increased follicular steroidogenesis in the presence and absence of TNF (28). However, the stimulatory effects of TNF on progesterone production were due to increased basal steroid production in the absence of exogenous steroid.

Time Course of Steroidogenesis Induced by TNF

Follicular progesterone production increased in response to TNF after 12 and 24 h of culture in either 5% CO_2 and air or 5% CO_2 and 95% O_2 (28). Androstenedione was low at 6 and 12 h and increased in all groups at 24 h. TNF stimulated only androstenedione production in the presence of 5% CO_2 and 95% O_2. TNF did not stimulate estradiol production in the presence of 5% CO_2 and air; in fact, estradiol was less than 1 ng in the 5% CO_2 and air group. In contrast, in 5% CO_2 and 95% O_2, follicles produced significantly greater basal amounts of estradiol compared to the 5% CO_2 and air group. Also, TNF increased estradiol above controls at 12 and 24 h in the 5% CO_2 and 95% O_2 group.

Because TNF increased estradiol under the conditions of 5% CO_2 and 95%

O_2, the effect of TNF on aromatase was ascertained (28). Addition of androstenedione alone significantly increased estradiol by follicles cultured under either high or low oxygen; however, in the presence of high oxygen, aromatase activity was greater than that in air. TNF in combination with androstenedione did not further stimulate estradiol above that of androstenedione alone under both culture conditions.

Steroid Production by Cultured Rat Preovulatory Theca

TNF increased progesterone, 20α-dihydroprogesterone, 17α-hydroxyprogesterone, and androstenedione at 24 h of culture in 5% CO_2 and air (28). In contrast, TNF stimulated progesterone, 20α-dihydroprogesterone, and 17α-hydroxyprogesterone but not androstenedione in the 5% CO_2 and 95% O_2. TNF had no significant effect on thecal testosterone or estradiol in both conditions.

Progesterone Production by Rat Granulosa Cells

Using standard conditions of 5% CO_2 and air, TNF inhibited granulosal progesterone at 24 h compared to granulosa cells cultured in the absence of TNF (28). FSH increased granulosal progesterone production, and this was slightly but significantly reduced by TNF.

Mechanism of TNF Action

In order to study the duration that TNF was required to stimulate progesterone production in vitro, preovulatory rat follicles were exposed to TNF during the first 1 or 6 h or during the entire 24-h period (These observations have not been published in a peer-reviewed journal by the authors.) Follicles exposed to TNF for 1 or 6 h showed progesterone levels no different than controls without TNF at these times. Exposure to TNF for 1 h produced progesterone levels after 24 h of culture that were no different than controls. However, incubation for 6 or 24 h in the presence of TNF produced progesterone levels after 24 h of culture that were significantly higher than controls.

To determine whether the action of TNF was related to increased levels of cAMP, follicles were incubated in the presence of 50 units TNF/mL for short (5–20 min) or long (6–24 h) periods, and cAMP levels were measured in media and follicles. cAMP levels were no different than control groups at any time period. In addition, when follicles were incubated with a higher amount of TNF (500 units/mL), cAMP levels in follicles after 24 h of culture were no different than either the control or 50-units-TNF/mL groups (control, 0.09 ± 0.02 ng cAMP/follicle; 500-units TNF, 0.14 ± 0.05). However, LH addition at two concentrations (5 ng or 5 μg/mL) increased cAMP levels in media and follicles 5–100 times higher than any other groups.

Nordihydroguaiaretic acid (NDGA) was added to culture medium to inhibit lipoxygenase activity. After 24 h of culture, progesterone accumulation in media was no different when follicles were incubated with TNF in the presence or absence of 3.3-μM NDGA. Similarly, when indomethacin (10 μM) was added to inhibit

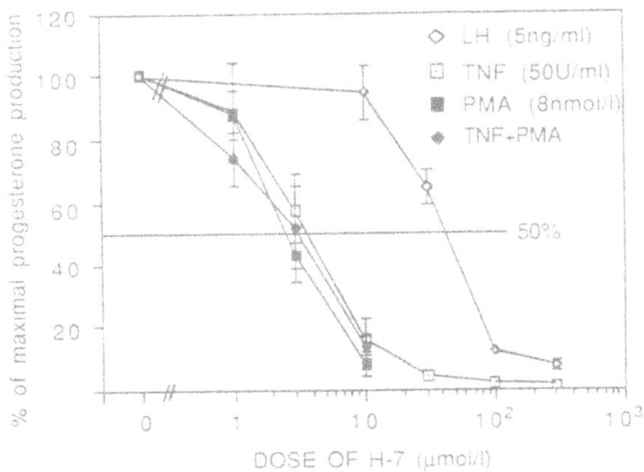

Fig. 5. Effect of protein kinase C inhibitor, H-7, on TNF-, PMA-, and LH-induced progesterone production. Two preovulatory rat follicles were incubated in 1 mL of incubation media in the presence of TNF, PMA, or LH and different amounts of H-7, and progesterone levels were measured in media after 24 h of culture. Values are expressed as percentage of maximal progesterone production of follicles not exposed to H-7.

cycloxygenase activity, TNF-stimulated progesterone was not altered. Also, PGE_2, $PGF_{2\alpha}$, and 6-keto-$PGF_{1\alpha}$ levels measured in media after 24 h of TNF stimulation revealed both PGE_2 and $PGF_{2\alpha}$ levels as undetectable (<10 pg/mL/2 follicles/24 h), and 6-keto-$PGF_{1\alpha}$ levels were detectable but no different than in TNF and control groups.

TNF-induced progesterone production was blocked by H-7, a PKC inhibitor, in a dose-dependent manner (Fig. 5). Inhibition of 50% corresponded to 5.2-µM H-7. LH-stimulated progesterone production was also inhibited by H-7 in a dose-dependent manner, but higher concentrations of H-7 were needed to inhibit LH action (50% inhibition corresponded to 54.5-µM H-7).

Phorbol 12-myristate-13 acetate (PMA), a PKC activator, increased progesterone in media in a dose-dependent manner, from 0- to 32-nM PMA. Follicles incubated with both TNF and PMA revealed progesterone levels that were further increased above PMA alone in an additive manner.

H-7 decreased progesterone production induced by PMA and PMA + TNF. A 50% decrease corresponded to 4.2- and 4.1-µM H-7 for PMA and PMA + TNF, respectively.

The calcium ionophore, A23187, increases intracellular calcium (by opening calcium membrane channels), which is necessary for PKC activation. A23187 induced a biphasic response of progesterone accumulation, with a maximum increase at 0.2-µM A23187. In the presence of 0.2-µM (or greater) A23187, there was no significant effect of TNF on progesterone levels in media.

When an extracellular calcium chelator, EGTA, was added to the follicular cultures, a dose-dependent increase of progesterone accumulation in media was observed, from 0- to 2-mM EGTA. TNF further increased progesterone in the presence of EGTA (2-way ANOVA, $P \leq 0.054$); however at 1- and 2-mM concentrations EGTA, TNF was unable to stimulate progesterone production.

Cycloheximide blocked TNF induced progesterone accumulation in media; while cycloheximide alone had no significant effect on basal progesterone production.

Immature Rats

A study by Emoto and Baird showed that TNF inhibited FSH-induced aromatase activity in a concentration-dependent fashion (29). At 10 ng/mL, TNF abolished the FSH-induced aromatase activity. A time course study revealed that TNF did not reduce aromatase activity per se, but prevented its induction by FSH. TNF could inhibit the induction of aromatase activity induced by both FSH and TGFβ. TNF also decreased progesterone synthesis that was stimulated by FSH. TGFβ (1 ng/mL) increased progesterone synthesis that was stimulated by FSH. The effects of FSH and TGFβ were blocked by TNF.

A study by Adashi, et al. also revealed that TNF could effectively and reversibly inhibit FSH-induced aromatase in the cultured granulosa cells of the DES-treated rat; a minimal time requirement of 48-h exposure to TNF was needed for inhibition (30). cAMP-generating agonists such as PGE_2, choleragen, and VIP also stimulated aromatase, and their actions could be attenuated by TNF. TNF did not inhibit FSH binding, but was capable of blocking forskolin-induced accumulation of cAMP and reducing adenylate cyclase activity. In other experiments TNF inhibited the ability of FSH to increase extracellular cAMP.

TNF also inhibited the production of progesterone and 20α-dihydroprogesterone while stimulating degradation of these steroids to 5α-pregnanediol (30). TNF alone slightly stimulated an increase in $[S^{35}]$sulfate incorporation into extracellular proteoglycans, but TNF significantly inhibited FSH-increased proteoglycan biosynthesis.

A portion of the results of Adashi, et al. (30) and Emoto and Baird (29) have recently been confirmed and extended (31). TNFα inhibited in a dose-dependent manner LH receptor formation induced by FSH, cholera toxin, forskolin, or 8-Br-cAMP; for FSH this action was observed as early as 24 h after TNF exposure and was due to a decrease in number of LH receptors, not affinity (31). As shown by Adashi, et al., TNF inhibited the cAMP production induced by FSH (30), and it also inhibited FSH-induced progesterone secretion (29–30).

Human

The following are unpublished observations of the authors: hCG stimulated progesterone production by granulosa-luteal cells in a time- and dose-dependent manner (Fig. 6). Progesterone production in the absence of hCG increased on days 2 and 4 of culture, then declined to a basal level throughout the remainder of the 10-day culture. Addition of hCG (0.05–1 IU/mL) increased progesterone production

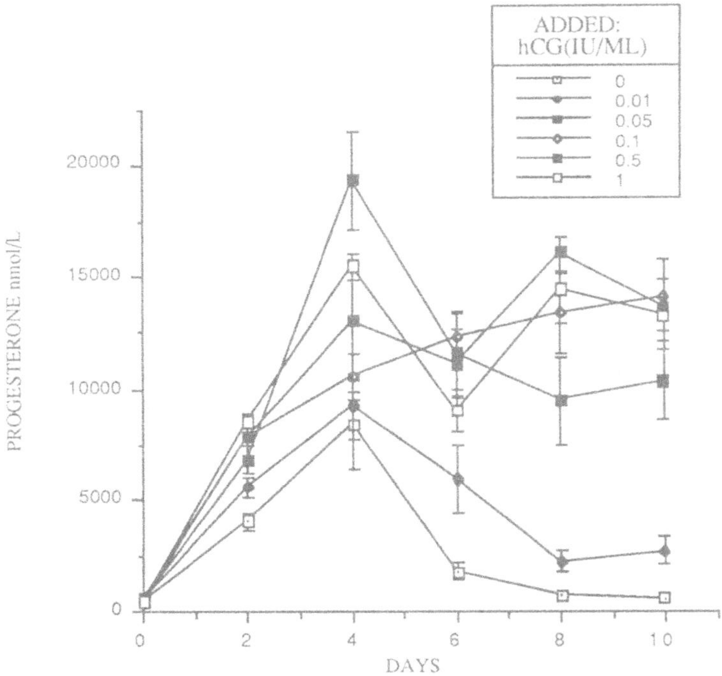

Fig. 6. Progesterone production by granulosa cells cultured in the presence of increasing concentrations of hCG. The media were changed, fresh media added, and the progesterone concentrations in the media were determined by a radioimmunoassay every 2 days for a 10-day culture period.

for the first 4 days and then plateaued throughout the remaining 10 days. The plateau phase was not maintained with the lowest dose (0.01 IU) of hCG tested, but it was above control levels. Progesterone was maximally stimulated with doses of 0.05- to 1-IU hCG during the 6th–10th days of culture.

Effect of TNF (10 ng/mL) and hCG (0.1 IU/mL) on Granulosa-Lutein Cell Progesterone Production

TNF significantly increased progesterone production above nonstimulated control cells by day 10 of culture (Fig. 7), and hCG increased progesterone above controls on days 2–10. The hCG-stimulated progesterone production increased steadily until day 4 of culture, and thereafter it remained unchanged. TNF and hCG in combination closely followed the pattern of progesterone production stimulated by hCG alone for the first 2–4 days. However, progesterone concentration continued to

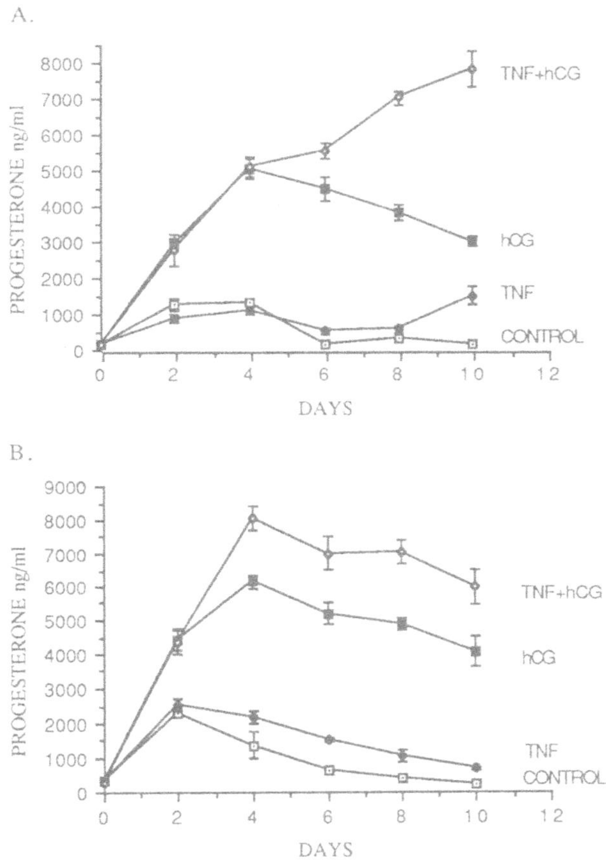

Fig. 7. The effect of TNF, hCG, and TNF plus hCG on granulosal progesterone production in two patients (A and B). Granulosa cells were cultured in the presence of TNF (0.6 nM), hCG (0.1 IU/mL), the combination of TNF and hCG, or media alone (control) for 10 days. The media were changed, and progesterone in the media was determined every 2 days by radioimmunoassay. $P < 0.001$ 2-way ANOVA, for the interaction of time and treatment. Patients A and B are representative of five patients examined.

increase for the remainder of the 10-day culture period under the influence of TNF and hCG.

Effect of TNF on hCG Binding

On day 10 of culture, the specific binding of hCG to granulosa cells was significantly higher in cells treated with TNF alone and hCG alone than controls. hCG binding in the group treated with TNF alone was significantly greater than the hCG group. hCG binding in the TNF group was not statistically different from the combination of TNF and hCG. However, hCG binding in the hCG plus TNF groups was significantly greater than that stimulated by hCG alone.

SUMMARY

In summary, the salient features of this overview are

1. Granulosa cells from antral follicles appear to be the source of ovarian TNFα.
2. Immunoreactive TNF appears in the healthy follicle at the time of antrum formation in the rat and human.
3. TNF appears to be quantitatively greater in atretic follicles.
4. In healthy preovulatory follicles of the rat, TNF stimulates androgen production.
5. Theca and granulosa are primary targets of TNF. In vitro TNFα increases thecal progesterone production and inhibits basal and FSH-stimulated granulosal progesterone from preovulatory follicles of the rat.
6. TNFα enhances hCG-stimulated progesterone production by granulosa-luteal cells in the human. This mechanism involves an increase in hCG binding induced by TNFα.
7. TNF in the human may be an autocrine and/or paracrine luteotropin because of its ability to synergize with hCG in stimulating progesterone production.

REFERENCES

1. Krishna A, Terranova PF. Alterations in mast cell degranulation and ovarian histamine in the proestrus hamster. Biol Reprod 1985;76:23-9.
2. Nakamura Y, Smith M, Krishna A, Terranova PF. Increased number of mast cells in the dominant follicle of the cow: Relationships among luteal, stromal, and hilar regions. Biol Reprod 1987;37:546-9.
3. Jiaswal K, Krishna A, Pandey LK. Human ovarian mast cells: Distribution and degranulation pattern. Proc XVII annu conf Endocr Soc India, Varanasi, 1987:20.
4. Krishna A, Beesley K, Terranova PF. Histamine, mast cells and ovarian function. J Endocrinol 1989;120:363-71.
5. Schreirber AB, Kenney J, Kowalski WJ, Friesel R, Mehlam T, Maciag T. Interaction of endothelial cell growth factor with heparin: Characterization by receptor and antibody recognition. Proc Natl Acad Sci 1985;82:6138-42.
6. Gospodarowicz D, Cheng J, Lui GM, Baird A, Esch F, Bohlen P. Corpus luteum angiogenic factor is related to fibroblast growth factor. Endocrinology 1985;117:2383-91.
7. Roby KF, Terranova PF. Stimulators and inhibitors of endothelial cell growth from the nondominant bovine ovary: Correlation to heparin binding. Biol Reprod 1987;36:87.
8. Roby KF, Terranova PF. Partial purification and characterization of an endothelial cell growth regulator from the bovine ovary. J Reprod Fertil 1990 (in press).

9. Sato N, Goto T, Haranaka K, et al. Actions of tumor necrosis factor on cultured endothelial cells: Morphologic modulation, growth inhibition, and cytotoxicity. J Natl Cancer Inst 1986;76:1113-21.
10. Schweigerer L, Malerstein B, Gospodarowicz D. Tumor necrosis factor inhibits the proliferation of cultured capillary endothelial cells. Biochem Biophys Res Commun 1987;143:997-1004.
11. Sugarman BJ, Aggarwal BB, Hass PE, Figari IS, Palladino MA Jr, Shepard HM. Recombinant human tumor necrosis factor-α: effects on proliferation of normal and transformed cells in vitro. Science (NY) 1985;230:943-5.
12. Vilcek J, Palombella VJ, Henrikson-DeStefano D, et al. Fibroblast growth enhancing activity of tumor necrosis factor and its relationship to other polypeptide growth factors. 1986;163:632-43.
13. Vilcek J, Tsujimoto M, Palombella VJ, Kohase M, Le J. Tumor necrosis factor: Receptor binding and mitogenic action in fibroblasts. J Cell Physiol 1987;5:57-61.
14. Fisch H, Gifford GE. In vitro production of rabbit macrophage tumor cell cytotoxin. Int J Cancer 1983;32:105-12.
15. Pennica D, Nedwin GE, Hayflick JS, et al. Human tumor necrosis factor: precursor structure, expression and homology to lymphotoxin. Nature (Lond) 1984;312:724-9.
16. Aggarwal BB, Kohr WJ, Hass PE, et al. Human tumor necrosis factor production, purification and characterization. J Biol Chem 1985;260:2345-4.
17. Old LJ. Tumor necrosis factor (TNF). Science (NY) 1985;230:630-2.
18. Smith RA, Baglioni C. The active form of tumor necrosis factor is a trimer. J Biol Chem 1987;262:6951-4.
19. Folkman J, Sullivan R, Butterfield C, Murray J, Klagsbrun M. Heparin affinity: Purification of a tumor-derived capillary endothelial cell growth factor. Science (NY) 1984;223:1296-9.
20. Beach RL, Popiela H, Festoff BW. The identification of neurotropic factor as a transferrin. FEBS Lett 1983;156:151-6.
21. Matthews N. Tumor necrosis factor from the rabbit, II. Production by monocytes. Br J Cancer 1978;38:310-5.
22. Matthews N, Watkins JF. Tumor necrosis factor from the rabbit, I. Mode of action, specificity and physicochemical properties. Br J Cancer 1978;38:302-9.
23. Bringman TS, Aggarwal BB. Monoclonal antibodies to human tumor necrosis factor alpha and beta: Application for affinity purification, immunoassays, and as structural probes. Hybridoma 1987;6:489-507.
24. Roby KF, Terranova PF. Localization of tumor necrosis factor (TNF) in rat and bovine ovary using immunocytochemistry and cell blot: Evidence for granulosal production. In: Hirshfield AN, ed. Growth factors and the ovary. New York: Plenum Press, 1989:273-8.
25. Roby KF, Weed J, Lyles R, Terranova PF. Immunologic evidence for a human ovarian tumor necrosis factor alpha. J Clin Endocrinol Metab 1990 (in press).
26. Bagavandoss P, Kunkel SL, Wiggins RC, Keyes PL. Tumor necrosis factor α (TNFα) production and localization of macrophages and T lymphocytes in the rabbit corpus luteum. Endocrinology 1988;122:1185-7.
27. Roby KF, Terranova PF. Tumor necrosis factor alpha alters follicular steroidogenesis in vitro. Endocrinology 1988;123:2952-4.
28. Roby KF, Terranova PF. Effects of tumor necrosis factor alpha in vitro on steroidogenesis of healthy and atretic follicles of the rat: Theca as a target. Endocrinology 1990;126.

29. Emoto N, Baird A. The effect of tumor necrosis factor/cachectin on follicle-stimulating hormone-induced aromatase activity in cultured rat granulosa cells. Biochem Biophys Res Commun 1988;153:792-8.
30. Adashi EY, Resnick CE, Croft CS, Payne DW. Tumor necrosis factor α inhibits gonadotropin hormonal action in nontransformed ovarian granulosa cells. J Biol Chem 1989;264:11591-7.
31. Darbon JM, Oury F, Laredo J, Bayard F. Tumor necrosis factor-α inhibits follicle-stimulating hormone-induced differentiation in cultured rat granulosa cells. Biochem Biophys Res Comm 1989;163:1038-46.

6

Transforming Growth Factors and Ovarian Function

David W. Schomberg and George W. Mulheron

Departments of Obstetrics and Gynecology, Cell Biology, and The Comprehensive Cancer Center, Duke University Medical Center, Durham, North Carolina

Transforming growth factor α (TGFα) is a 50-amino acid polypeptide that interacts with the same receptor as the epidermal growth factor (EGF) (1). It is, therefore, a regulatory molecule capable of affecting the same cell types, including epithelial cells, that respond to EGF (2). Transforming growth factor βs (TGFβs), on the other hand, are 25-kD peptides consisting of two identical 112-amino acid subunits that are synthesized as part of a larger, biologically latent peptide (3–4). They are considered to be part of a more extensive gene family that includes other ovarian effectors, such as inhibin, activin, and mullerian inhibiting substance (5). The cellular effects of TGFβs are multifactorial and are dependent upon other factors in the milieu, but in general, their effect upon epithelial cells is antimitotic (3). Two excellent reviews of the biochemical properties and cellular actions of TGFα and TGFβs are presented elsewhere in this volume.

With respect to the ovary, TGFα and $TGF\beta_1$ have been demonstrated in vitro to modulate the function of cells representing every compartment of the ovarian follicle (i.e., oocyte-cumulus, granulosa, and theca-interstitium) (6–7). In the context of growth factor physiology (i.e., local action), this array of effects implies that TGFs are autocrine and/or paracrine effectors produced within the follicle. In terms of endocrine control of follicle growth and differentiation, the next

Acknowledgments: Supported in part by Grants HD-11827, HD-21261, and HD-07315 from the NICHD, NIH. We want to thank Ms. R.R. Newbold, National Institute of Environmental Health Sciences, Research Triangle Park, NC, for her invaluable assistance and expertise in the preparation of the photomicrographs. We wish to acknowledge the gift of oFSH from the NIDDK and the NHPP, University of Maryland School of Medicine. The authors wish to recognize the excellent technical contributions of Ms. Amanda Carver and Ms. Askale Mathias. The expert secretarial assistance of Ms. Pam Vowell is also most gratefully acknowledged.

step is to determine the extent to which their production is regulated by reproductive hormones. Evidence is beginning to accumulate that indicates that the TGFs or TGF-like compounds exemplify these considerations. The most recent findings of this type are discussed here, the vast majority of which have become available since the publication of several chapters on the role of TGFs in the ovary (6, 8–10).

ATTEMPTS TO DEMONSTRATE TGF ACTIVITY IN VIVO

Previous Studies

Gospodarowicz, et al. performed the initial experiments designed to demonstrate effects in vivo of EGF and fibroblast growth factor (FGF) upon ovarian follicular development (11). The growth factors were administered intraperitoneally (5 μg/day for 5 days) to 10- to 13-day-old rats, and the ovaries examined histologically. Ovarian area was increased approximately 1.4-fold by EGF, indicating the development/maintenance of a larger number of follicles, but no specific acceleration of development or atresia in a defined population of follicles was reported. To our knowledge, in the only other published study utilizing rodents, EGF (about 4 μg/animal) was administered daily to newborn female mice for the first 5 days of life (12). EGF significantly reduced the number of growing follicles at more advanced stages of development and decreased the percentage of mitotic figures in the follicular and surface epithelia of the ovary. In postpubertal sheep a depilatory dose of EGF (4.2 μg/kg/h) for 48 h decreased LH and estradiol concentrations and delayed the onset of next estrus; serum-FSH concentrations were increased (13). In a more detailed study, Radford, et al. found that a similar EGF infusion inhibited ovulation and lowered estrogen concentrations in PMSG-treated ewes; subsequent nonstimulated cycles were apparently normal (14).

Current Studies

We designed studies in the prepubertal rat to evaluate the effect of locally administered growth factors. EGF or TGFβ_1 was administered directly into one ovary, while the contralateral ovary received vehicle only; FSH was given subcutaneously to some animals. In the first study, 21-day-old intact females received a DES-containing silastic implant. On day 25 they were assigned to the following experimental subgroups: (*a*) No FSH; (*b*) FSH (NIH S-17), 7.5, 2.5, or 0.75 μg/injection in phosphate-buffered saline (PBS), every 12 h for 5 injections; and (*c*) Metrodin (Serono), 0.15, 0.05, or 0.015 units/injection in PBS similarly administered. Within each subgroup, one ovary received randomly either 25-ng TGFβ_1 or 250-ng EGF in 3% methyl cellulose in PBS as a sequestering agent; the contralateral ovary received vehicle only. Additional method controls were performed with 10-, 25-, and 100-ng TGFβ_1, or 250- and 1000-ng EGF in 3% methyl cellulose or in PBS in the absence of FSH. EGF was dissolved in PBS, pH 7.8, and TGFβ was dissolved in 4-mM HCl, then diluted further with PBS, pH 7.8. Animals were anesthetized with ketamine intraperitoneally (IP) (50 mg/mL, 0.105 cc/100 g body weight), and the ovaries exteriorized. With the aid of a magnifying glass, 10 μL of solution was delivered via Hamilton syringe into the interior of the ovary. Twelve hours after completion of the FSH injections, the animals were sacrificed, the ovaries were fixed in 4% paraformaldehyde-PBS, and histological staining was performed with hematoxylin and eosin. Serial sections were obtained utilizing up to 50% of the ovary.

The expected dose-responses to FSH were obtained, and the degree of stimulation at the 7.5-μg dose was comparable to that described previously (15). The magnitude of follicular stimulation obtained with FSH and Metrodin indicated the two preparations were comparable on the basis of their bioactivity. No obvious differences were apparent between growth factor-injected and control ovaries, either in the absence of FSH or in its presence at any dose. Although a statistical morphometric analysis was not performed, the results of the method control study suggested that the 3% methyl cellulose vehicle actually enhanced the degree of atresia, perhaps secondary to carrier-induced release of inhibitory cytokines and/or growth inhibitory factors. Another methodologic observation concerned the injection volume; 10 μL was unsatisfactory because a few instances of leakage occurred. In summary, these studies showed that EGF or TGFβ_1 administered directly into the ovary in the doses utilized did not produce significant changes in follicular morphology.

A second study was carried out using hypophysectomized, 25-day-old females following the same injection schedules outlined above. In this study 0.75-μg FSH and 0.015-U Metrodin were used because these doses produced a very uniform degree of early antral stimulation. Since the ovaries in this model are considerably smaller, the growth factor doses (100-ng EGF; 100-ng TGFβ_1) were delivered in 2-uL PBS. Methyl cellulose carrier was not used. In another subgroup of animals, the approach of Gospodarowicz (11) was utilized, and 100-ng TGFβ_1 or 1.0- and 10.0-μg EGF were administered IP for each of 3 days. Also in this subgroup the bursa of one ovary was randomly cut open to determine whether this procedure might provide easier access of the IP-injected material to the ovaries and thereby enhance any responses.

As in the previous study, the follicular response to FSH and Metrodin was comparable. Again, no significant morphological differences were apparent between control and growth factor-injected ovaries. Likewise, no consistent effect upon follicular morphology was apparent, either in the presence or absence of FSH or Metrodin in the subgroup that received IP-administered growth factors. As shown in the photomicrographs, ovarian size differed slightly, but no significant differences were noted in terms of enhanced development or atresia at any specific stage of follicle development (Figs. 1–3). The results of the method control portion of this study (bursotomy vs. intact bursa) suggested that surgical insult to the bursa resulted in an increased degree of follicular atresia, again suggesting that the release of cytokines or growth inhibitory factors might be causing local changes. In summary, in the hypophysectomized, non-DES-treated model, neither intraovarian nor intraperitoneal administration of EGF or TGFβ_1 in the doses utilized elicited significant developmental or atretic changes in follicular morphology.

Our inability to demonstrate effects in vivo upon ovarian morphology should not be interpreted to indicate that TGFs do not play a significant role in

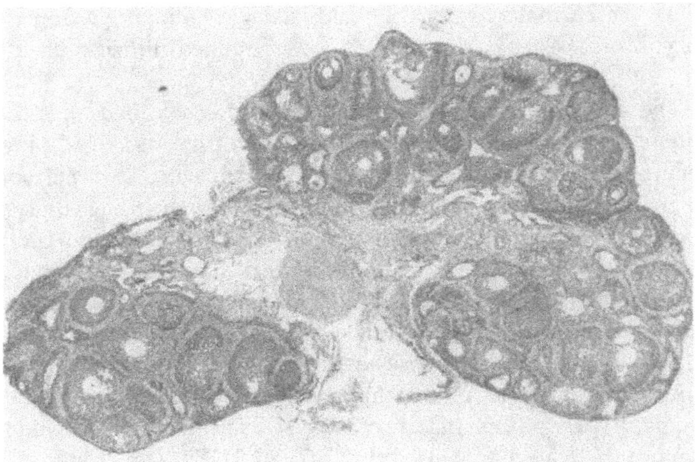

Fig. 1. Ovary of a control hypophysectomized animal in which FSH treatment was begun on day 25. The ovary is representative of those that received 0.75-μg oFSH or 0.015 U Metrodin every 12 h for 5 injections. See text for additional details.

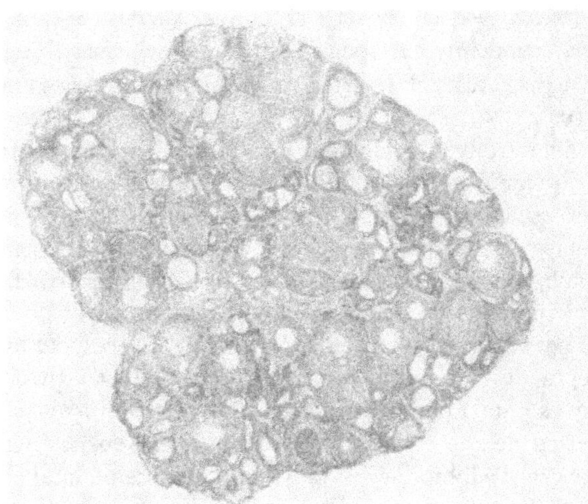

Fig. 2. Ovary from an animal treated with FSH as in Figure 1 that concurrently received 100-ng $TGF\beta_1$ IP for 3 days. No significant differences in follicle development or atresia relative to the control are apparent. See text for additional details.

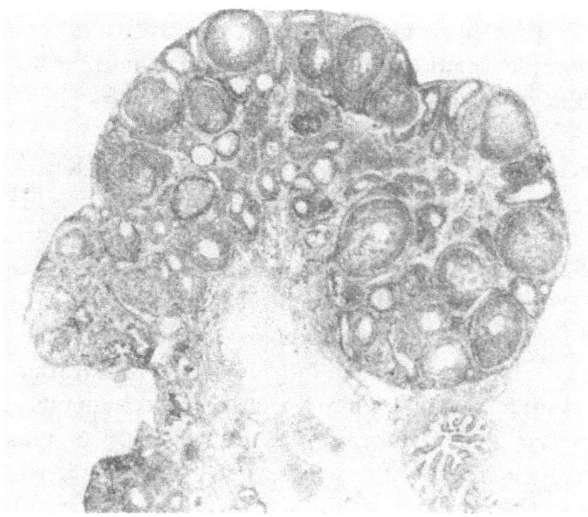

Fig. 3. Ovary from an animal treated with FSH as in Figure 1 that concurrently received 10-μg EGF IP for 3 days. No significant differences in follicular development or atresia relative to the control are apparent. See text for additional details.

ovarian physiology. Several caveats should be kept in mind with respect to this particular experimental approach: (*a*) The end point was much less sensitive than specific biochemical indices, (*b*) the doses utilized may have been too small, (*c*) the growth factors may have rapidly diffused away from the site of injection, and (*d*) the ovarian "test stages" (DES treatment or 4-days' post hypophysectomy) may not have been optimal. In any event, these results do call for different in vivo models. In the rat, refinements of the approach could be attempted vis à vis the caveats mentioned. Alternatively, ovaries of prepubertal domestic animals might provide a superior model because their larger size would facilitate accurate and reproducible intraovarian placement of larger doses. The ultimate approach will be to utilize molecular biological approaches to induce hyper- or hypo-expression of growth factor genes specifically in the ovary and to observe the consequences of these manipulations.

FURTHER DEMONSTRATIONS OF TGF ACTIVITY IN VITRO

Recent studies, which collectively encompassed all the various compartments of the ovarian follicle, have extended the range of cellular responses modulated by TGFα/EGF and/or TGFβ. Using cumulus-enclosed mouse oocytes maintained in meiotic arrest with hypoxanthine, Downs tested the ability of a number of growth factors to induce germinal vesicle breakdown (16). Of those tested, only EGF stimulated a high frequency of germinal vesicle breakdown. Also, EGF was the only growth-promoting factor that triggered cumulus cell expansion. In contrast, TGFβ did not significantly affect either response. These results and others led to the conclusion that a specificity exists for EGF-like molecules in the maturation of the oocyte as well as the cumulus oophorus. Further analyses showed that hormone- or growth factor-stimulated cumulus expansion and oocyte maturation in vitro were not causally related.

In contrast to these results in the mouse model, TGFβ accelerated spontaneous oocyte maturation in rat oocytes; it also suppressed the positive effect of EGF (17). In another study utilizing the rat model, however, treatment with TGFα alone, like LH, induced the resumption of meiosis in follicle-enclosed oocytes (18). In this study TGFβ did not significantly affect spontaneous oocyte maturation, but partially suppressed LH-induced resumption of meiosis. Collectively, these studies demonstrate clearly that TGFs can influence murine oocyte maturation and emphasize the importance of understanding more about their mechanisms of action.

In a related aspect of the ovulatory process, the tissue-type plasminogen activator (tPA) is specifically increased after exposure to gonadotropins. Like FSH, EGF and TGFα stimulated the secretion of tPA activity. Furthermore, EGF/TGFα stimulated tPA mRNA levels in cultured rat granulosa cells by pathways independent of protein kinases A and C (19). Further studies are needed to elucidate the molecular mechanisms involved, but these results provide another example of the involvement of these growth factors in important ovarian cellular processes.

In another ovarian cell type, the germinal epithelium, FGF is an autocrine stimulator of growth, at least in vitro. In this cell type, TGFβ inhibits FGF-stimulated growth, providing another example of the ability of TGFβ to modulate the bioactivity of FGF (20). These FGF-TGFβ interactions may well be important in the regulation of follicle development prior to their gonadotropin-dependent stages.

The thecal-interstitial cell (TIC) is also a target of TGFβ in terms of androgenic function in vitro. In rat TICs, TGFβ acts directly to suppress androgen production (21). TGFβ, alone or in combination with LH, stimulated an increase in cholesterol side-chain cleavage ($P450_{scc}$), but did not alter cellular 17α hydroxylase/C17-20 lyase content. Under these conditions the TIC became primarily progesterone-producing cells, suggesting that TGFβ may alter the ability of C17-20 lyase to act on its substrate (21). In thecal cells from large porcine follicles, TGFβ inhibited basal and gonadotropin-stimulated secretion of progesterone, androstenedione, and testosterone, but stimulated basal and gonadotropin-induced production of estradiol (22). EGF did not alter thecal secretion of progesterone, androstenedione, or testosterone, but significantly inhibited basal and hCG-stimulated secretion of estradiol. These results emphasize again the fact that growth factors definitely have roles as local modulators of steroidogenesis, but the precise, specific responses that they elicit are dependent upon species and/or the developmental state of the follicle. The biochemical reasons for these differences are not known.

Species comparisons with respect to granulosa cell steroidogenesis extend to the human. EGF caused a dose-dependent inhibition of FSH-induced aromatase activity in gonadotropin-stimulated granulosa cells obtained as part of an in vitro fertilization program, but had little effect on activity in the absence of FSH. EGF also markedly inhibited FSH-stimulated cytochrome P450 aromatase mRNA levels in these cells (23). These results agree with those found previously using cultured rat granulosa cells (24).

Another species comparison highlights the fact that cultured rat granulosa cells have never demonstrated significant mitotic capability in vitro relative to those of the pig, cow, and human (25). However, Dorrington, et al. demonstrated mitotic responsiveness utilizing a combination of TGFβ and FSH (26). The maximum obtainable response was a 4-fold increase in cell number over a 7-day period. This is not a sustained response typical of a growth factor-stimulated system, as can be accomplished with porcine granulosa cell cultures and EGF, for example (27). Hirshfield and Schmidt argue that such a limited response is to be expected since rat granulosa cells of this developmental stage (stage VIII follicles) are essentially end-stage cells nearing the end of their replicative potential, and they have been programmed to begin their terminally differentiated, highly steroidogenic phase in response to gonadotropins (28). Nonetheless, this FSH-TGFβ interaction is of considerable interest and may involve regulation of the c-*myc* and/or c-*fos* oncogenes, the expression of which is stimulated by FSH in rat granulosa cells (29).

Lastly, in the granulosa cell model, EGF/TGFα and TGFβ have been demonstrated to modulate the production of other growth factors, specifically immunoreactive IGF-I (iIGF-I). In cultured porcine granulosa cells, EGF/TGFα increase production of iIGF-I, and this level of production is attenuated by TGFβ (30–31). These observations point to another level of complexity in the system wherein the local production of growth factors may be modulated not only by the classical reproductive hormones, but also by the growth factors themselves.

PRODUCTION AND REGULATION OF TGFs IN OVARIAN CELLS

EGF/TGFα

As reviewed previously, a potential route by which growth factors and reproductive hormones could interact to control reproductive function is via the action of reproductive hormones to modulate the cellular production of growth factors. This paradigm is currently being demonstrated in the ovary in terms of IGF-I and the TGFs. The experimental evidence demonstrating modulation of ovarian IGF-I mRNA expression and protein secretion is presented elsewhere in this volume. Evidence for modulation of TGF production has been both direct and indirect. TGFα mRNA expression or TGFα secretion have been demonstrated in the theca of the rat and the cow (32–34). FSH stimulation of message, but not of product, was reported in the rat (35). Developmental expression was demonstrated in bovine theca (36). In this study only antral follicles 0.7–2.0 mm in diameter showed intense TGFα immunostaining, thus implying very precise regulation of production. In the hamster ovary, EGF-like activity has been localized immunohistochemically (37). This activity is very closely related to EGF immunologically; it did not crossreact with TGFα, IGF-I, or FGF. Staining activity varied with the stage of the estrous cycle. Furthermore, expression was controlled by gonadotropins; staining intensity after long-term hypophysectomy was increased especially by FSH.

TGFβ

With the identification of mRNAs for three (TGFβ_1, β_2, and β_3) of the five known forms of TGFβ in cells of the mammalian ovary (see Flanders, et al., this volume), the potential role(s) of the TGFβs as intraovarian regulatory molecules has been expanded. Consequently, regulation of the intraovarian expression of these TGFβs should be examined. What follows is a synopsis of both the in vivo and in vitro studies that address the question of TGFβ expression by ovarian cells.

In vivo FSH treatment of prepubertal rats stimulated whole-ovary TGFβ_1 expression in a preliminary report (38). In light of our observation that granulosa cells do not express detectable TGFβ_1 mRNA by Northern analysis using a cDNA probe (see below), presumably some FSH-initiated signal modified TGFβ_1 mRNA expression in the nongranulosa compartment(s). Utilizing an immunocytochemical approach with TGFβ_1-specific antibodies, Thompson, et al. provided evidence for thecal and interstitial localization of TGFβ_1 in vivo, but did not detect TGFβ_1 in the granulosa compartment (39). On the other hand, using more sensitive techniques, expression of TGFβ_1 mRNA was reported preliminarily in isolated GC, but levels of the transcripts were not altered by FSH (40). Studies of cycling rats in vivo showed that TGFβ_1 mRNA was low in preovulatory follicles, but was increased within 12 h by an ovulatory dose of hCG; message level was maintained in developing CL and during the CL of pregnancy (41). More in-depth study of both GC and thecal-interstitial responses in whole-ovary and isolated cell systems will be necessary to understand these various interrelationships.

In vitro, primary cultures of bovine theca/interstitial cells produced higher levels of TGFβ activity than did granulosa cell (GC) cultures (42). In contrast, we recently found considerable TGFβ-like activity in the conditioned culture medium of rat granulosa cells grown on a fibronectin substratum (43). The subtype of TGFβ involved was not identified in either study. Given the immunocytochemical results of Thompson, et al. (38), we postulated that the TGFβ activity found in rat GC-conditioned culture medium was TGFβ type 2. This idea was initially tested by Northern analysis of total RNA isolated from cultured rat GC using cDNA probes for TGFβ_1 and TGFβ_2. The TGFβ_2 cDNA hybridized to two bands: a major species approximately 5.1 kb and a minor 3.6-kb species. TGFβ1 mRNA was not detectable in granulosa cell RNA. Both TGFβ_2 and TGFβ_1 mRNA species were detectable in control MCF-7 cell RNA. Furthermore, FSH significantly reduced the relative expression of TGFβ_2 mRNA by cultured cells, as well as the level of TGFβ-like activity in the culture medium (44). In current studies, the nature of the TGFβ-like activity in the medium was investigated using Western blot analysis and was shown to be TGFβ_2 (Fig. 4).

These results suggest that TGFβ_2 is at least one type of TGFβ that is hormonally regulatable in granulosa cells. In at least one other cell type (i.e., primary cultures of rat keratinocytes), TGFβ_2 rather than TGFβ_1 is the subtype that is regulatable; its expression/production is increased by retionic acid (45). Relative levels of TGFβ_1 mRNA are increased by estrogen in human osteosarcoma cells (46), and levels of TGFβ_2 mRNA in prostatic adenocarcinoma cells are increased by

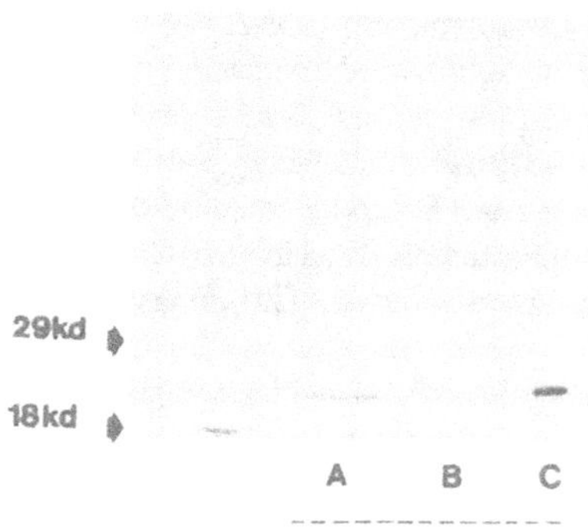

Fig. 4. Western blot of rat GC-conditioned culture medium demonstrating the presence of TGFβ_2. Following SDS-PAGE and transfer to nitrocellulose, the filter was incubated with anti-TGFβ_2 antibody (R&D Systems, Minneapolis, MN) followed by a peroxidase-conjugated anti-IgG second antibody. (Lane A, conditioned medium; lane B, 25-ng standard TGFβ_1; lane C, 25-ng standard TGFβ_2.)

tamoxifen (47). Thus, the regulation of TGFβ_1/β_2 mRNA expression or secretion of the active gene product may vary with the cell type. While to our knowledge there have been no studies on the regulation of TGFβ_3 expression in the ovary, it has been detected immunohistochemically in uterine epithelium; the intensity of staining was modified by steroids (Flanders, et al., this volume). Therefore, because both steroids and a gonadotropin have been demonstrated to modulate mRNA expression or production of the various TGFβs, a plausible working hypothesis with respect to hormonal regulation of granulosa cell growth and differentiation would be that TGFβ_2 mRNA is normally expressed in a developmentally dependent fashion, with the level of expression being altered by steroid and/or gonadotropic hormones. Further testing of this hypothesis utilizing various approaches, including in situ hybridization and immunocytochemistry, will establish whether TGFs can, in fact, be considered as a class of intermediate effectors acting in the ovary between the classical reproductive hormones and their target cells.

REFERENCES

1. Marquardt H, Rose TM, Webb NR, et al. Rat transforming growth factor type I: Structure and relation to epidermal growth factor. Science 1984;223:1079-82.
2. Todaro GJ, Fryling C, DeLarco JE. Tranforming growth factors produced by certain human tumor cells: Polypeptides that interact with epidermal growth factor receptors. Proc Natl Acad Sci USA 1980;77:5158-62.
3. Sporn MB, Roberts AB, Wakefield LM, Assoian RK. Transforming growth factor-β: Biological function and chemical structure. Science 1986;233:532-40.
4. Wakefield LM, Smith DL, Flanders KC, Sporn MB. Latent transforming growth factor-β from human platelets. J Biol Chem 1988;263:7646-54.

5. Mason AJ, Hayflick JS, Ling N, et al. Complementary DNA sequences of ovarian follicular fluid inhibin show precursor structure and homology with tranforming growth factor-β. Nature 1985;318:659-63.
6. Knecht M, Feng P, Catt KJ. Transforming growth factor-beta: Autocrine, paracrine, and endocrine effects in ovarian cells. Seminars Reprod Endocrinol 1989;7:12-20.
7. Schomberg DW. Growth factors and reproduction. In: Hodgen GD, Rosenwaks Z, Spieler JM, eds. Nonsteroidal gonadal factors. Norfolk, CT: Jones Institute Press, 1988:330-8.
8. May JV, Schomberg DW. The potential relevance of epidermal growth factor and transforming growth factor-alpha to ovarian physiology. Seminars Reprod Endocrinol 1989;7:1-11.
9. Skinner MK. Transforming growth factor production and action in the ovarian follicle: Thecal cell-granulosa cell interactions. In: Hirshfield AN, ed. Growth factors and the ovary. New York: Plenum Press, 1989:141-50.
10. Hammond JM. Paracrine regulation in the ovary. In: Medically assisted conception, an agenda for research. Institute of Medicine, National Academy Press, 1989:211-33.
11. Gospodarowicz D, Mescher AL, Birdwell CR. Control of cellular proliferation by the fibroblast and epidermal growth factors. In: Third decennial review conference: Cell tissue and organ culture. National Cancer Institute Monograph 48. Bethesda, MD: USPHS, NIH, 1978:109-130.
12. Lintern-Moore S, Moore GPM, Panaretto BA, Robertson D. Follicular development in the neonatal mouse ovary: Effect of epidermal growth factor. Acta Endocrinol (Copenh) 1981;96:123-6.
13. Shaw G, Jorgenson G, Tweendale R, Tennison M, Waters MJ. Effect of epidermal growth factor on reproductive function of ewes. J Endocrinol 1985;107:429-36.
14. Radford HM, Avenell JA, Panaretto BA, Some effects of epidermal growth factor on reproductive function in merino sheep. J Reprod Fertil 1987;80:113-8.
15. Richards JS. Estradiol receptor content in rat granulosa cells during follicular development: Modification by estradiol and gonadotropins. Endocrinology 1975;97:74-84.
16. Downs SM. Specificity of epidermal growth factor action on maturation of the murine oocyte and cumulus oophorus in vitro. Biol Reprod 1989;41:371-9.
17. Feng P, Catt KJ, Knecht M. Transforming growth factor-β stimulates meiotic maturation of the rat oocyte. Endocrinology 1988;122:181-8.
18. Tsafrirri A, Vale W, Hsueh AJW. Effects of transforming growth factors and inhibin-related proteins on rat preovulatory graafian follicles in vitro. Endocrinology 1989; 125:1857-62.
19. Galaway AB, Oikawa M, Ny T, Hsueh AJW. Epidermal growth factor stimulates tissue plasminogen activator activity and messenger ribonucleic acid levels in cultured rat granulosa cells: Mediation by pathways independent of protein kinases A and C. Endocrinology 1989;125:126-35.
20. Gospodarowicz D, Plouet J, Fugii DK. Ovarian germinal epithelial cells repond to basic fibroblast growth factor and express its gene: Implications for early folliculogenesis. Endocrinology 1989;125:1266-76.
21. Magoffin DA, Gancedo B, Erickson GF. Transforming growth factor-β promotes differentiation of ovarian thecal-interstitial cells but inhibits androgen production. Endocrinology 1989;125:1951-8.
22. Caubo B, DeVinna RS, Tonetta SA. Regulation of steroidogenesis in cultured porcine granulosa cells by growth factors. Endocrinology 1989;125:321-6.

23. Mendelson CR, Means GD, Mahendroo MS, et al. Use of molecular probes to study regulation of aromatase cytochrome P-450. Biol Reprod 1990;42:1-10.
24. Hsueh AJW, Adashi EY, Jones PCB, Walsh TH Jr. Hormonal regulation of the differentiation of cultured ovarian granulosa cells. Endocr Rev 1984;5:76-127.
25. Gospodarowicz D, Bialecki H. Fibroblast and epidermal growth factors are mitogenic for cultured granulosa cells of rodent, porcine, and human origin. Endocrinology 1979;104:757-64.
26. Dorrington J, Chuma AV, Bendall JJ. Transforming growth factor β and follicle-stimulating hormone promote rat granulosa cell proliferation. Endocrinology 1988;123:353-9.
27. Gospodarowicz D. The control of proliferation of ovarian cells by the epidermal and fibroblast growth factors. In: Spilman CH, Wilks JW, eds. Novel aspects of reproductive physiology. New York: SP Medical and Scientific Books, 1978:107-80.
28. Hirshfield AN, Schmidt WA. Kinetic aspects of follicular development in the rat. In: Mahesh VB, Dhindsa DS, Anderson E, Kalra SP, eds. New York: Plenum Press, 1987; 211-36.
29. Delidow BC, White BA, Peluso JJ. Gonadotropin induction of c-fos and c-myc expression and deoxyribonucleic acid synthesis in rat granulosa cells. Endocrinology 1990;126:2302-6.
30. Mondschein JS, Hammond JM. Growth factors regulate immunoreactive insulin-like growth factor-I production by cultured porcine granulosa cells. Endocrinology 1988; 123:463-8.
31. Mondschein JS, Canning SF, Hammond JM. Effects of transforming growth factor-β on the production of immunoreactive insulin-like growth factor I and progesterone and on [^{3}H-T]thymidine incorporation in porcine granulosa cell cultures. Endocrinology 1988;123:1970-6.
32. Skinner MK, Coffey RJ Jr. Regulation of ovarian cell growth through the local production of transforming growth factor-α by thecal cells. Endocrinology 1988; 123:2632-8.
33. Lobb DK, Skinner MK, Dorrington JH. Rat thecal/interstitial cells produce a mitogenic activity that promotes the growth of granulosa cells. Mol Cell Endocrinol 1988; 55:209-17.
34. Bendell JJ, Lobb DK, Chuma A, Gysler M, Dorrington JH. Bovine thecal cells secrete a factor(s) that promotes granulosa cell proliferation. Biol Reprod 1988;38:79-97.
35. Kudlow JE, Korbin MS, Purchio AF, et al. Ovarian transforming growth factor-α gene expression: Immunohistochemical localization to the thecal/interstitial cells. Endocrinology 1987;121:1577-9.
36. Lobb DK, Korbin MS, Kudlow JE, Dorrington JH. Transforming growth factor-alpha in the adult bovine ovary: Identification in growing ovarian follicles. Biol Reprod 1989; 40:1087-93.
37. Roy SK, Greenwald GS. Immunohistochemical localization of epidermal growth factor-like activity in the hamster ovary with a polyclonal antibody. Endocrinology 1990;126:1309-17.
38. Hernandez ER, Twardzik DR, Purchio A, Adashi EY. Gonadotropin-dependent ovarian transforming growth factor-β gene expression [Abstract 35]. Biol Reprod 1987;1 (suppl).
39. Thompson NL, Flanders KC, Smith JM, Ellingsworth LR, Roberts AB, Sporn MB. Expression of transforming growth factor-β1 in specific cells and tissues of adult and

neonatal mice. J Cell Biol 1989;108:661-9.
40. Knecht M, Feng P, Catt K, Gelmann E. Expression of the TGF-β gene in the ovary and its role during meiosis of rat oocytes [Abstract D113]. J Cell Biochem 1988.
41. Gaddy-Kurten D, Hickey GJ, Fey GH, Gouldie J, Richards JS. Hormonal regulation and tissue-specific localization of α-2 macroglobulin in rat ovarian follicles and corpora lutea. Endocrinology 1989;125:2985-95.
42. Skinner MK, Keski-Oja J, Osteen KG, Moses HL. Ovarian thecal cells produce transforming growth factor-β which can regulate granulosa cell growth. Endocrinology 1987;121:786-92.
43. Kim I-C, Schomberg DW. The production of transforming growth factor-β activity by rat granulosa cells. Endocrinology 1989; 124:1345-51.
44. Mulheron GW, Schomberg DW. Rat granulosa cells express transforming growth factor-β type 2 messenger ribonucleic acid which is regulatable by follicle-stimulating hormone in vitro. Endocrinology 1990;126:1777-9.
45. Glick AB, Flanders KC, Danielpour D, Yuspa SH, Sporn MB. Retionic acid induces transforming growth factor-β2 in cultured keratinocytes and mouse epidermis. Cell Regulation 1989;1:87-97.
46. Komm BS, Terpening CM, Benz DJ, et al. Estrogen binding, receptor mRNA, and biologic response in osteoblast-like osteosarcoma cells. Science 1988;241:81-4.
47. Ikeda T, Toubin MN, Marquardt H. Human transforming growth factor type B2; production by a prostatic adenocarcinoma cell line, purification and initial characterization. Biochemistry 1987;26:4337-45.

7

Intraovarian IGF-I System

Eli Y. Adashi,[1] Carol E. Resnick,[1] Eleuterio R. Hernandez,[1] Arye Hurwitz,[1] Charles T. Roberts,[2] Derek Leroith,[2] and Ron Rosenfeld[3]

[1]Division of Reproductive Endocrinology, Department of Obstetrics and Gynecology, University of Maryland School of Medicine, Baltimore; [2]NIH, NIDDK, Bethesda, Maryland; and [3]Pediatric Endocrinology, Stanford University Medical Center, Stanford, California

As the significance of putative intraovarian regulators becomes increasingly recognized, much of the attention centers on insulin-like growth factors (IGFs). Indeed, a large body of evidence now suggests the existence of an intraovarian IGF system complete with ligands, receptors, and binding proteins. More importantly, IGFs have been shown to exert a variety of significant effects at the level of the somatic ovarian cell, raising the possibility of a meaningful in vivo role. The above notwithstanding, there is at this time no compelling evidence to indicate that IGFs (or for that matter any other putative intraovarian regulator) are indispensable to ovarian function. However, a large body of a somewhat indirect nature strongly suggests such a possibility. It is the purpose of this communication to review and summarize key developments in this area.

OVARIAN IGF-I PRODUCTION

Ovarian production of IGF-I was initially suggested by studies revealing that the immunoreactive (i) IGF-I content of porcine follicular fluid substantially exceeded that encountered in serum (1). Further evidence consisted of the demonstration of cycloheximide-inhibitable, gonadotropin-dependent (and estradiol-dependent) iIGF-I in serum-free media conditioned by cultured porcine granulosa cells (2–3). Although the rat ovarian content of iIGF-I appears to be growth hormone-dependent (4), a direct effect of growth hormone at the level of the rat granulosa cell remains to be demonstrated.

Acknowledgments: Supported in part by NIH Research Grant HD-19998 and USPHS Research Career Development Award 1-K04-HD-00697 from the NICHHD, NIH.

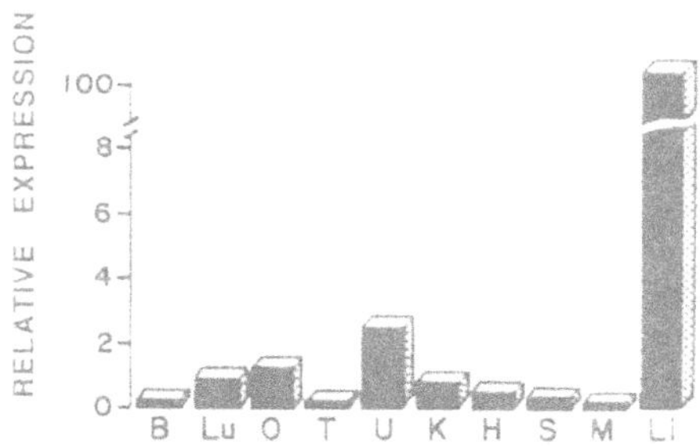

Fig. 1. IGF-I gene expression: Tissue distribution. (From Murphy LJ, Bell GI, Friesen HB, Tissue distribution of insulin-like growth factors I and II messenger ribonucleic acid in the adult rat, Endocrinology 1987;120:1279.)

Extending the investigation to the transcriptional level, we have recently shown that the adult and immature rat ovary (as well as the isolated, immature granulosa cell) is a site of IGF-I gene expression and that it may be subject to gonadotropic regulation (5). Significantly, of all adult rat organs tested (6), the ovary (O) displays the third-highest level of IGF-I gene expression, the uterus (U) and liver (Li) being the most active in this regard (Fig. 1). In contrast, human granulosa cells may be a site of IGF-II, rather than IGF-I gene expression (7). These observations and the lack of IGF-II gene expression in the adult rat ovary (6) suggest possible species specificity. However, profound differences in the experimental conditions may favor IGF-II gene expression as a dedifferentiation (fetal) marker.

We have recently undertaken to assess the relative ovarian abundance of IGF-I transcripts with alternative 5'-untranslated (UT) regions, their cellular localization, and hormonal regulation (5). To this end, a solution hybridization/RNase protection assay was employed, wherein total rat ovarian RNA was hybridized with a 404-base P-labeled rat IGF-I riboprobe corresponding to the class A 5'-UT variant. As in liver, three protected bands [322 (class A), 297 (class B), and 242 (class C) bases long] were noted, in keeping with established alternative 5'-UT transcripts. The ovarian (as the hepatic) class C variant proved the most abundant. The ovarian class B variant was barely detectable.

Cellular localization studies revealed these ovarian IGF-I transcripts to be primarily, if not exclusively, of granulosa but not thecal-interstitial cell origin (Fig. 2). Treatment of immature (21–23 days old) hypophysectomized rats with a subcutaneous silastic implant containing diethylstilbestrol (DES) for a total of 5 days resulted in a 2-fold increase in the (densitometrically quantified) abundance of ovarian IGF-I transcripts, a diametrically opposed effect (2.6-fold decrease) being noted at the level of the liver. Whereas treatment of hypophysectomized rats with oGH by itself (150 μg, q.i.d., sc, ×5 days) resulted in a 5-fold increase in hepatic IGF-I gene expression, a limited, albeit distinct, inhibitory effect was observed on the steady state levels of ovarian IGF-I mRNA. In contrast, combined treatment with

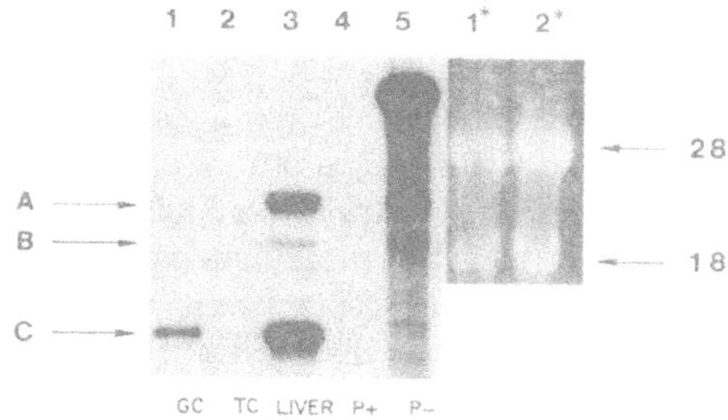

Fig. 2. Ovarian IGF-I gene expression: Cellular localization. (GC = granulosa cells, TC = thecal-interstitial cells.)

oGH and DES yielded a 3-fold increase in the abundance of ovarian IGF-I transcripts, there being no net alteration in hepatic IGF-I gene expression. Taken together, these findings reveal ovarian expression of the three known 5'-UT IGF-I mRNA variants, document the granulosa cell as the main somatic ovarian cell of IGF-I mRNA generation, and indicate that hepatic and ovarian IGF-I gene expression are differentially regulated in diametrically opposed directions.

These and related studies have resulted in a tentative consensus as to the regulation of ovarian (or more specifically, granulosa cell) IGF-I gene expression. According to this view, granulosa cell-derived IGF-I biosynthesis may well be growth hormone-, FSH-, and estrogen-dependent. It is further presumed that granulosa cell-derived IGF-I is capable of traversing the cellular plasma membrane into the extracellular fluid, wherein it may bind to specific cell-membrane type I IGF receptors on its very cell of origin or else on adjacent cells. In doing so, granulosa cell-derived IGF-I may exert autocrine or paracrine effects, respectively. These concepts are indeed at the very heart of the intraovarian IGF hypothesis.

Stated differently, it has been presumed that locally generated IGF-I would, in fact, play a meaningful role in vivo by regulating its cell of origin or else immediately adjacent cell types. Although an attractive possibility, this notion has thus far not received rigorous experimental verification. Preliminary studies would suggest, however, that immunoneutralization of granulosa cell-derived IGF-I may inhibit, at least in part, FSH- and growth hormone-mediated granulosa cell differentiation. The above notwithstanding, much additional work would be required before unequivocal conclusions could be reached as to the in vivo relevance of IGF-I to ovarian physiology. Indeed, limitations imposed by current experimental approaches may well require that definite proof await more sophisticated experimental paradigms. In particular, it is to be hoped that transgenic technology would develop to a level sufficient to allow selective ablation of IGF-I at the level of the ovary such that reproductive function could be evaluated in both the experimental group and the controls.

OVARIAN IGF-I RECEPTION

Both porcine (8) and murine (9) granulosa cells have now been shown to display high-affinity, low-capacity binding sites for IGF-I. In our hands (10), binding to FSH-primed rat granulosa cells proved time-, temperature-, and pH-dependent, optimal steady state conditions being achieved following an 8-h incubation at 15°C and a pH of 8.0. Although subject to regulation by the cellular density of plating, the binding of [^{125}I]IGF-I to its receptor proved saturable (apparent Kd = 3.3×10^{-9}M), as well as reversible, complete, or partial tracer displacement being effected by competitive inhibition and dilution, respectively. Scatchard and Hill analyses yielded linear plots consistent with a single class of noninteracting binding sites. Specificity studies revealed the competition for [^{125}I]IGF-I binding to follow a rank order of potency of IGF-I > MSA > insulin, a pattern compatible with a type I IGF receptor. Limited or no displacement was observed for a series of chemically related and unrelated polypeptides, as well as by a human insulin receptor antiserum. Using affinity crosslinking, we have also been able to observe that whole ovarian membranes of untreated (or FSH-treated) immature, hypophysectomized, DES-treated rats are endowed with specific type I IGF receptors. Similar results have later been obtained with isolated granulosa cells from the same experimental model (unpublished).

We have recently reported FSH and LH to be capable of up-regulating granulosa cell IGF-I binding, an effect further augmented by growth hormone, but not prolactin (11). Specifically, we were able to show that FSH is capable of up-regulating IGF-I binding in a time- and dose-dependent fashion and that cAMP, its purported intracellular second messenger, may play an intermediary role in this regard. Indeed, granulosa cell IGF-I binding was enhanced following elevation of the intracellular cAMP content by a series of cAMP-generating agonists, inhibition of cAMP-phosphodiesterase activity, or the provision of nondegradable cAMP analogs. High-dose forskolin (10^{-5}M), like FSH, proved capable of augmenting IGF-I binding by itself, while an essentially inert dose (10^{-7}M) synergized with FSH in this regard. Significantly, heterologous receptor up-regulation was not limited to FSH, similar increments being observed for luteotropic, β_2-adrenergic, but not lactogenic, granulosa cell agonists. Related in vivo studies using immature, hypophysectomized, DES-treated rats revealed that the ability of FSH to up-regulate granulosa cell IGF-I binding (*a*) is not strictly an in vitro phenomenon and can be fully reproduced in vivo; (*b*) is due to enhancement of IGF-I binding capacity rather than affinity; (*c*) may be subject to diametrically opposed modulation by somatogenic and GnRH-like granulosa cell agonists (up- and down-regulation, respectively); and (*d*) is best maintained by gonadotropins, but not prolactin. Inasmuch as gonadotropin dependence constitutes a unique attribute of the ovarian granulosa cell, our findings further suggest that the granulosa cell IGF-I receptor may have thoroughly adapted to its unique environment, providing the first example of a cell type for which the complement of IGF-I receptors may be cAMP dependent.

Given the pivotal role of FSH in the induction of granulosa cell receptors for luteotropic and lactogenic ligands, this finding strongly suggests that the acquisition of IGF-I responsiveness may be part and parcel of granulosa cell ontogeny. Accordingly, gonadotropins may condition the cell to respond optimally to IGF-I, thereby conferring selective advantage upon follicles so endowed.

Both insulin-like growth factor (IGF-I) and IGF-II have been shown to promote granulosa cell differentiation and proliferation. While both type I and type II IGF receptors have been observed in rat granulosa cells, the identity of the IGF receptor type(s) mediating IGF receptor action cannot be completely evaluated at this time due to the lack of specific reagents. The availability of antibodies specific for the rat type II IGF receptor (R-II-PAB_1) has made studies of this receptor type possible. To validate the utility of the R-II-PAB_1 antiserum at the level of the rat granulosa cell, its ability to immunoneutralize the granulosa cell type II IGF receptor was examined (12). Significantly, R-II-PAB_1 (10–100 μg/mL) proved a potent inhibitor of [^{125}I]IGF-II (but not [^{125}I]IGF-I) binding to granulosa cell membrane preparations. Substantial, albeit finite R-II-PAB_1-mediated inhibition of the crosslinking of [^{125}I)IGF-II was also observed. Moreover, R-II-PAB_1 proved highly potent in immunoprecipitating the rat granulosa cell type II IGF receptor.

In light of these observations, we have proceeded to use R-II-PAB_1 to assess the functional role of the rat granulosa cell type II IGF receptor in IGF-I and IGF-II hormonal action. To this end, 20-ng/mL-FSH-primed granulosa cells were cultured for 72 h in the absence or presence of IGF-I or IGF-II (50 ng/mL), with or without increasing (receptor-active) concentrations of R-II-PAB_1 (10–100 μg/mL). Control incubations were carried out with an ammonium sulfate precipitate of nonimmune rabbit serum dialyzed against phosphate-buffered saline. Significantly, both R-II-PAB_1 and nonimmune rabbit serum were without effect on the cytodifferentiative action of either IGF-I or IGF-II. Subject to limitations inherent to the immunoneutralizing potency of R-II-PAB_1, these findings are in keeping with the notion that (inasmuch as the conventional cytodifferentiative process is concerned) the granulosa cell type II IGF receptor does not appear to participate in transmembrane IGF signaling. By inference, these findings also suggest that IGF-I and IGF-II hormonal action at the level of the granulosa cell may be exerted largely, if not exclusively, via the type I IGF receptor. Thus, the potential relevance and the functional role(s), if any, of the granulosa cell type II IGF receptor remain to be determined.

GRANULOSA CELL AS A SITE OF IGF-I ACTION

Studies carried out in the last several years have clearly established the granulosa cell as a site of IGF-I action. IGF-I action at the level of the rat, but not porcine, granulosa cell appears largely (but not exclusively) contingent upon its ability to synergize with pituitary gonadotropins (Fig. 3). These effects are unaccounted for by enhanced cellular viability, plating efficiency, or DNA synthesis. Thus, this ability of IGF-I to augment differentiated phenotypic expression of the developing granulosa cell may be distinct from its well-established growth-promoting property and was thus considered a novel biologic effect of this polypeptide (13). In this

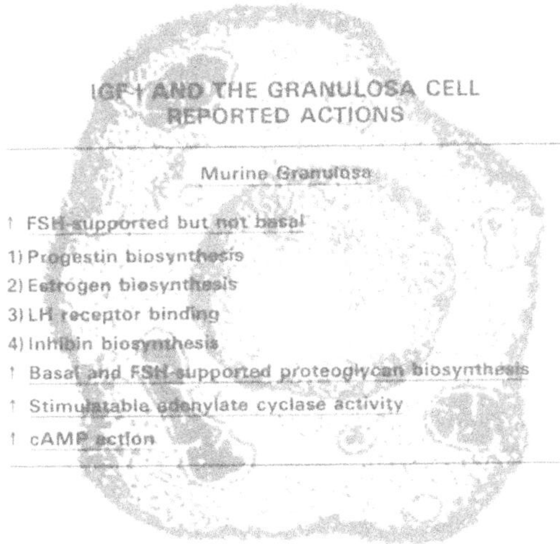

Fig. 3. IGF-I and the murine granulosa cell: Reported actions.

connection, we have been able to show that IGF-I is capable of augmenting FSH-supported (but not basal) progesterone and estrogen biosynthesis, as well as the FSH-mediated acquisition of LH receptors.

More recently, IGF-I was also found to augment basal as well as FSH-supported proteoglycan biosynthesis (14). In this respect, IGF-I appeared to exert its classic "sulfation factor" activity at the level of the granulosa cell, the very chondrotropic effect that led to its discovery. Fractionation of the major extracellular proteoglycan species revealed FSH to favor the exclusive production of dermatan sulfate, whereas IGF-I supported simultaneous biosynthesis of both heparin and dermatan sulfate. These findings suggest that IGF-I may effect marked quantitative as well as qualitative alterations in proteoglycan economy. Given the possible role of proteoglycans in follicular antrum formation and follicular atresia, these findings raised the possibility that IGF-I of granulosa cell origin may partake in the growth as well as demise of the developing ovarian follicle.

GRANULOSA CELL AS A SITE OF IGF-BINDING PROTEIN GENERATION

Insulin-like growth factor (IGF) binding proteins (IBPs) are multifunctional proteins that regulate not only the transport of IBPs, but also their presentation to cell surface receptors. The latter function is thought to be subserved by growth hormone-independent low-molecular weight IBPs capable of binding IGFs (but no insulin) with affinities in the range of 10^{-10} to 10^{-9}M (15–16). One such murine binding moiety of hepatocyte (BRL-3A cell line) origin (IBP_2) has recently been purified (17–18) and cloned (18–19), revealing a 270-residue (29.5-kD) mature, nonglycosylated protein endowed with an RGD adhesion (cell attachment) sequence (20), commonly (but not exclusively) observed in matrix proteins. Although the precise role(s) of tissue-derived IBPs remains a matter of study, both stimulation (21–22) and inhibition (23–24) of IGF binding and action have been reported. Thus, the synthesis and secretion of IBPs may play a major role in the regulation of IGF hormonal action at the target cell level, even at a time when the extracellular concentrations of the IGFs remain constant. As such, this development adds a new level of complexity to the interaction of IGFs at the cellular level.

Although the relevance of IBPs to ovarian physiology remains unknown, immunoreactive IBP_1 (previously referred to as placental protein-12) has been found in abundance in human ovarian follicular (25) and cyst (26) fluid. Although not detected in unstimulated ovarian tissue or in nonluteinized granulosa cells, IBP_1 was localized to luteinized human granulosa cells, as well as to corpora lutea (25). Moreover, metabolically labeled (highly luteinized) human granulosa cells derived from ovaries subjected to gonadotropic hyperstimulation have been found capable of de novo synthesis of an IBP_1-related species as assessed by immunoisolation (27). In related experiments acid-chromatographed whole-ovarian extracts of murine origin (28), as well as media conditioned by porcine granulosa cells (2–3), were found to contain hormonally dependent, low-molecular weight IGF-I binding activity. Most recently, IBP_3 of porcine follicular fluid origin was found to suppress FSH hormonal action at the level of the murine granulosa cell (29).

To explore the possibility that the granulosa cell is also capable of hormonally regulatable elaboration of IBPs, granulosa cells from immature, DES-primed rats were cultured for up to 72 h under serum-free conditions in the absence or presence of FSH (30). Media conditioned by untreated granulosa cells reveal constitutively released polyethylene glycole-precipitable [^{125}I]IGF-I binding activity, the daily elaboration of which proved constant throughout the 72 h experimental period. However, treatment of granulosa cells with FSH (Fig. 4) resulted in dramatic inhibition of the accumulation of IGF-I binding activity (89% at the 100-ng/mL dose level). Systemic provision of FSH (10 μg/rat/day for 2 days) revealed that this gonadotropic action is not strictly an in vitro phenomenon, but can be fully reproduced under in vivo circumstances. Western ligand blotting of SDS-PAGE-fractionated media conditioned by untreated granulosa cells revealed three IGF-BP species comprising a major band doublet (28–29 kD), as well as a single minor band (23 kD). Treatment with FSH virtually eliminated the 23-kD species and substantially reduced the relative representation of the 28- and 29-kD IGF-BP species (82% and 74% inhibition, respectively). Taken together, these observations disclose the multiplicity of granulosa cell-derived IGF-BPs and reveal the striking ability of FSH to suppress their constitutive release under both in vitro and in vivo circumstances. This FSH action is all the more noteworthy in light of the generally stimulatory effect exerted by FSH at the level of the granulosa cell. Inasmuch as FSH may be concerned with the promotion of granulosa cell development, its ability to attenuate the release of (presumptively inhibitory) IGF-BPs may enhance the

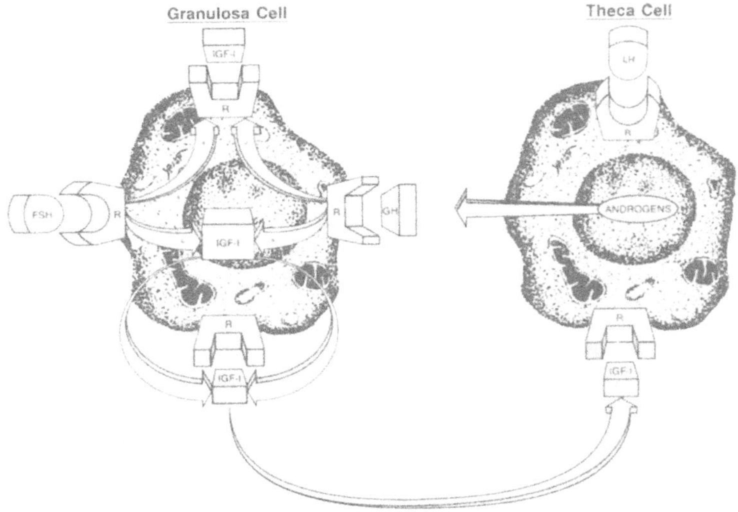

Fig. 4. FSH-attenuated elaboration of IGF-I binding activity: IBPs species. (BRL 3A = media conditioned by a liver cell line; serum = adult rat serum.)

access of endogenously produced IGF-I to its cognate cell surface receptors and, hence, its cellular hormonal action.

SUMMARY

Although much remains to be learned with respect to the possible relevance of IGF-I to ovarian physiology, it may be possible at this time to tentatively formulate possible functions of IGF-I in this connection: (*a*) *amplification* of gonadotropin hormonal action—a key requirement given the exponential nature of follicular development; (*b*) *integration* of follicular development—an essential facet concerned with the coordination of granulosa-theca cooperation (Fig. 5); and (*c*) *selection* of dominant follicle(s)—a speculative proposition assuming timely and selective activation of the IGF-I system in “chosen” follicles.

Aside from its possible role(s) in the course of established follicular cycles, IGF-I (and/or IGF-II) may also participate in the very formation of the follicular apparatus during the late fetal/early neonatal period. Although the ovary is gonadotropin-independent at that time, we have previously shown that IGF-I may well interact with VIPergic input now implicated in the morphodifferentiation of the follicular apparatus. Similarly, IGF-I may be concerned with the promotion of juvenile and early pubertal follicular gonadotropin (FSH) levels. Ovarian IGF-I may have a bearing on the puberty-promoting effect of growth hormone. Indeed, an association appears to exist between isolated growth hormone deficiency and

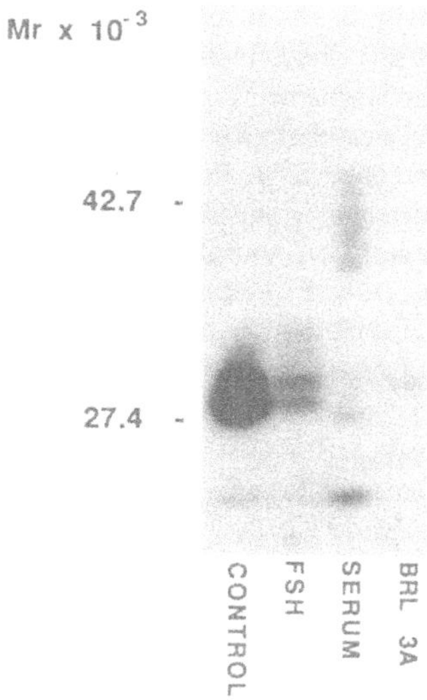

Fig. 5. Intraovarian intercompartmental interactions: Granulosa-theca interstitial cell coordination.

delayed puberty in both rodents and human subjects, a process reversed by systemic growth hormone replacement therapy. Given that ovarian IGF-I and its receptor may be growth hormone-dependent, it is tempting to speculate that the ability of growth hormone to accelerate pubertal maturation, may be due, at least in part, to the promotion of ovarian IGF-I production and reception with the consequent local potentiation of gonadotropin action (31).

REFERENCES

1. Hammond JM. Peptide regulators in the ovarian follicle. Aust J Biol Sci 1981;34: 491-504.
2. Hammond JM, Baranao JLS, Skaleris D, Knight AB, Romanus JA, Rechler MM. Production of insulin-like growth factors by ovarian granulosa cells. Endocrinology 1985;117:2553-5.
3. Hsu C-J, Hammond JM. Gonadotropins and estradiol stimulate immunoreactive insulin-like growth factor-I production by porcine granulosa cells in vitro. Endocrinology 1987;120:198-207.
4. Davoren JB, Hsueh AJW. Growth hormone increases ovarian levels of immunoreactive somatomedin-C/insulin-like growth factor I in vivo. Endocrinology 1986;118: 888-90.
5. Hernandez ER, Roberts CT, LeRoith D, Adashi EY. Rat ovarian insulin-like growth factor (IGF-I) gene expression is granulosa cell-selective: 5'-untranslated mRNA variant representation and hormonal regulation. Endocrinology 1989;125:572-4.

6. Murphy LJ, Bell GI, Friesen HB. Tissue distribution of insulin-like growth factor I and II messenger ribonucleic acid in the adult rat. Endocrinology 1987;120:1279-82.
7. Voutilainen R, Miller WL. Coordinate tropic hormone regulation of mRNAs for insulin-like growth factor II and the cholesterol side-chain-cleavage enzyme, P450scc, in human steroidogenic tissues. Proc Natl Acad of Sci USA 1987;84:1590-4.
8. Veldhuis JD, Furlanetto RW, Juchter D, Garmey J, Veldhuis P. Trophic actions of human somatomedin-C/insulin-like growth factor I on ovarian cells: In vitro studies with swine granulosa cells. Endocrinology 1985;116:1235-42.
9. Davoren JB, Kasson BG, Li CH, Hsueh AJW. Specific insulin-like growth factor (IGF) I- and II-binding sites on rat granulosa cells: Relation to IGF action. Endocrinology 1986;119:2155-62.
10. Adashi EY, Resnick CE, Hernandez ER, Svoboda ME, Van Wyk JJ. Characterization and regulation of a specific cell membrane receptor for somatomedinC/insulin-like growth factor I in cultured rat granulosa cells. Endocrinology 1988;122:194-201.
11. Adashi EY, Resnick CE, Svoboda ME, Van Wyk JJ. Follicle-stimulating hormone enhances somatomedin-C binding to cultured rat granulosa cells: Evidence for cAMP-dependence. J Biol Chem 1986;261:3923-6.
12. Adashi EY, Resnick CE, Rosenfeld RG. Insulin-like growth factor-I (IGF-I) and IGF-II hormonal action in cultured rat granulosa cells: Mediation via type I but not type II IGF receptors. Endocrinology 1989.
13. Adashi EY, Resnick CE, D'Ercole AJ, Svoboda ME, Van Wyk JJ. Insulin-like growth factors as intraovarian regulators of granulosa cell growth and function. Endocr Rev 1985;6:400-20.
14. Adashi EY, Resnick CE, Svoboda ME, Van Wyk JJ, Hascall VC, Yanagishita M. Independent and synergistic actions of somatomedin-C in the stimulation of proteoglycan biosynthesis by cultured rat granulosa cells. Endocrinology 1986;118:456.
15. Binoux M, Hossenlopp P, Hardouin S, Seurin D, Lassarre C, Gourmelen M. Somatomedin (insulin-like growth factors)-binding proteins: Molecular forms and regulation. Horm Res 1986;24:141-51.
16. Baxter RC. The insulin-like growth factors and their binding proteins. Comp Biochem Physiol 1988;91B:229-35.
17. Mottola C, MacDonald RG, Brackett JL, Mole JE, Anderson JK, Czeck MP. Purification and amino-terminal sequence of an insulin-like growth factor-binding protein secreted by rat liver BRL-3A cells. J Biol Chem 1986;261:11180-8.
18. Brown AL, Chiariotti L, Orlowski CC, et al. Nucleotide sequence and expression of a cDNA clone encoding a fetal rat binding protein for insulin-like growth factors. J Biol Chem 1989;264:5148-54.
19. Margot JB, Binkert C, Mary J-L, Landwehr J, Heinrich G, Schwander J. A low molecular weight insulin-like growth factor binding protein from rat: cDNA cloning and tissue distribution of its messenger RNA. Mol Endocrinol 1989;3:1053-66.
20. Ruoslahti E, Pierschbacher MD. New perspectives in cell adhesion: RGD and integrins. Science 1987;238:491-7.
21. De Vroede MA, Tseng LY-H, Katsoyannis PG, Nissley SP, Rechler MM. Modulation of insulin-like growth factor I binding to human fibroblast monolayer cultures by carrier proteins released to the incubation media. J Clin Invest 1986;77:602-13.
22. Elgin RG, Busby WJ Jr, Clemmons DR. An insulin-like growth factor (IGF) binding protein enhances the biologic response to IGF-I. Proc Natl Acad Sci USA 1987; 84:3254-8.

23. Drop SLS, Valiquette G, Guyda HJ, Corvol MT, Posner BI. Partial purification and characterization of a binding protein for insulin-like activity (ILAs) in human amniotic fluid: A possible inhibitor of insulin-like activity. Acta Endocrinol Copenh 1979; 90:505-18.
24. Ritvos O, Tanta P, Jalkanen J, et al. Insulin-like growth factor (IGF) binding protein from human decidua inhibits the binding and biological action of IGF-I in cultured choriocarcinoma cells. Endocrinology 1988;122:2150-7.
25. Seppala M, Wahlstrom T, Koskimies AI, et al. Human preovulatory follicular fluid, lutenized cells of hyperstimulated preovulatory follicles, and corpus luteum contain placental protein 12. J Clin Endocrinol Metab 1984;58:505-10.
26. Seppala M, Than G. Insulin-like growth factor binding protein PP12 in ovarian cyst fluid. Arch Gynecol Obstet 1987;241:33-5.
27. Suikkari AM, Jalkanen J, Koistinen R, et al. Human granulosa cells synthesize low molecular weight insulin-like growth factor-binding protein. Endocrinology 1989; 124:1088-90.
28. Davoren JB, Hsueh AJW. Growth hormone increases ovarian levels of immunoreactive somatomedin C/insulin-like growth factor I in vivo. Endocrinology 1986;118: 888-90.
29. Ui M, Shimonaka M, Shimasaki S, and Ling N. An insulin-like growth factor-binding protein in ovarian follicular fluid blocks follicle-stimulating steroid production by ovarian granulosa cells. Endocrinology 1989;125:912-6.
30. Adashi EY, Resnick CE, Hernandez ER, Hurwitz A, Rosenfeld RG. Follicle-stimulating hormone inhibits the constitutive release of insulin-like growth factor binding proteins by cultured rat ovarian granulosa cells. Endocrinology 1990 (in press).
31. Homburg R, Eshel A, Abdalla HI, Jacobs HS. Growth hormone facilitates ovulation induction by gonadotropins. Clin Endocrinol 1988;29:113-7.

PART

III

REGULATION OF NORMAL AND NEOPLASTIC MAMMARY GROWTH

8

EGF-Mediated Growth Control and Signal Transduction in the MDA-MB-468 Human Breast Cancer Cell Line

Ronald N. Buick

Ontario Cancer Institute and Department of Medical Biophysics, University of Toronto, Ontario, Canada

Epidermal growth factor (EGF) is a mitogenic polypeptide affecting proliferation of a variety of cells. The tissue-specific actions of EGF are mediated through binding to a transmembrane receptor glycoprotein (EGFR) composed of an extracellular EGF-binding domain, a transmembrane segment, and an intracellular domain that has tyrosine kinase activity (1). The EGFR has been subject to intense study in relation to cancer biology because of its homology to the transforming protein encoded by the avian oncogene V-*erb* B (2).

EGFR IN BREAST CANCER

Evidence for a role for EGF in growth control of mammary epithelial cells has come from a number of sources (Table 1), including the effect of EGF on development of the mammary gland during pregnancy (3), carcinogenesis of the breast in rodent systems (4), and an obligatory role in maintenance of primary breast epithelial cultures of rodent or human origin (5–6). EGF is also a mitogen for many breast cancer cell lines in tissue culture (7), and many breast tumors express high levels of the receptor protein as evidenced by EGF binding or immunohistochemical staining (8). It is of interest that there is an inverse relationship between such EGFR expression and estrogen receptor (ER) status; it has been suggested, therefore, that in ER-ve tumors the EGFR may be involved in growth control.

The mechanism underlying the overexpression of the EGFR in biopsies of breast tumors has not been elucidated in detail. However, unlike other tumor types that overexpress the EGFR, such as squamous carcinoma and glioblastoma (9–10), the predominant mechanism is not based on receptor gene amplification. Although approximately 20% of breast tumors express elevated levels of EGFR, only ~10% of these have EGFR gene amplification. Since the cell of origin of ER-ve breast tumors is not established, and the level of EGFR expression in differentiating

Table 1. EGF: Role in mammary epithelial development and carcinogenesis.

- It is necessary for rodent gland development during pregnancy (3).
- It stimulates proliferation and morphological development of rodent and human tissue in in vitro systems (5–6).
- It is involved in rodent spontaneous mammary tumor development (4).
- It is a mitogen for a number of breast cancer cell lines in culture (7).
- Level of EGFR in biopsies is related to prognosis in biopsies of human breast cancer; is overexpressed in ER-ve tumors (11).

normal mammary epithelium is not known, it remains a possibility that the variability in expression may be unrelated to any mutational events in the carcinogenesis process, but rather is secondary to the expansion of a rare cell population that normally expresses high levels of EGFR.

As an attempt to derive laboratory tissue culture models of the subclass of breast tumors overexpressing the EGFR, we screened a series of breast cancer cell lines for EGF-binding capacity. We identified the estrogen-receptor negative cell line MDA-MB-468 as expressing approximately 1 to 2 × 10^6 EGFR/cell, approximately 20- to 100-fold higher than levels expressed in other breast tumor cell lines or normal fibroblasts (12). In this case the overexpression is based on an amplification of the receptor gene; Southern blotting indicated a 20- to 40-fold amplification, and in situ hybridization located the amplified domain to an abnormally banding region on one copy of chromosome 7 (13). No evidence of structural rearrangements in the EGFR gene have been detected. In addition, the properties of the overexpressed protein appear normal with respect to binding affinity, synthesis, turnover, and autophosphorylation (13–14). The following discussion emphasizes the utility of this cell line in investigations of the molecular mechanisms underlying the involvement of the EGF receptor in breast cancer.

GENERATION OF MDA-MB-468 CELL VARIANTS

Cell lines overexpressing the EGFR commonly can be growth inhibited in tissue culture by exposure to supraphysiological levels of EGF. The mechanism of this phenomenon has not been established, but it has been clearly related to receptor frequency; using variants selected from A431 carcinoma cells expressing different levels of receptor, Kamamoto, et al. (15) demonstrated a "threshold" requirement in terms of receptor number to allow growth inhibition to occur, and in a series of squamous tumor cell lines, Kamata, et al. (16) showed a direct relationship between growth inhibition and increasing receptor frequency. Despite these relationships it is of interest that overexpression of the EGFR alone does not seem to confer the ability to be growth inhibited by high levels of EGF. In circumstances where expression of EGFR is imposed on normal fibroblasts after transfection with expression vectors, clones of transformed fibroblasts are generated that express high levels of EGFR, but that are not growth inhibited by supraphysiological levels of EGF (17–18).

In keeping with the experience with squamous carcinoma cells, MDA-MB-468 is also growth inhibited in tissue culture by high levels (10^{-7}M) of EGF (12). Clones selected for survival under these conditions were all found to express low numbers ~2 × 10^4/cell) of receptor. In 12 selected clones the mechanism used to overcome the EGF growth inhibition was loss of the copy of chromosome 7 bearing the amplified allele of the EGFR. This was demonstrated both by karyology and by the fact that the amplified allele of the EGFR in MDA-MB-468 cells has a useful restriction fragment length polymorphism (13). Such variants exist within the parental MDA-MB-468 cell line at a frequency of approximately $1/10^6$.

MDA-MB-468 VARIANTS

Six of the 12 subclones (S1, S4, S5, S10, S11, and S12) were selected for investigation of EGF-mediated growth control. In plastic-adherent tissue culture containing 10% fetal calf serum, all 6 clones could be moderately stimulated in terms of growth by the addition of EGF, and supraphysiological levels of EGF did not cause growth inhibition. When suspended in anchorage-independent conditions, clones S1, S5, S10, S11, and S12 displayed the same phenotype as in adherent culture, but clone S4 was dependent on the addition of exogenous EGF for growth (13). Similarly, when the tumorigenicity of the clones was assessed, S4 displayed a phenotype separate from the other 5 clones. Clones S1, S5, S10, S11, and S12 were all tumorigenic, forming progressively growing tumors after subcutaneous implantation of 10^6 cells in the flank or mammary fat pad of nude mice; clone S4, however, was nontumorigenic under the same conditions. When the rates of tumor growth of clones S1, S5, S10, S11, and S12 were compared with parental MDA-MB-468 cells, a growth rate advantage could be detected for the parental cell line (doubling time of 6 days vs 10 days).

The tumor growth experiments therefore suggest that amplification of the EGFR gene and subsequent overexpression of the receptor protein provide a growth advantage in vivo. Similar conclusions have been reached through study of the in vivo growth of A431 and variant cell lines (19). These experiments provide a biological rationale for the in vivo selection of tumor cell clones overexpressing the EGFR during tumor progression. Based on analysis of receptor expression in relation to tumor stage (11), overexpression of EGFR is likely a late event in tumorigenesis. Our data are consistent with that view in that the predominant class of subclones (S1, S5, S10, S11, and S12) are still tumorigenic and, therefore, must possess other derangements of cell growth regulatory mechanisms. One such abnormality was described recently: MDA-MB-468 cells have a homozygous deletion of the recessive oncogene RB1 that confers susceptibility to retinoblastoma and possibly also to other tumor types (20).

We have been unable, as yet, to explain the different growth properties of S4 and the predominant class of subclones. Two possible explanations for the differential response to EGF in agar culture are suggested by other work in the area

Table 2. Comparison of EGFR properties (*A*) and transcription of transforming growth factors genes (*B*) in MDA-MB-468 and subclones S4 and S5.

	468	S4	S5
A:			
EGFR Freq.	2×10	3.3×10^4	1.8×10^4
KD (M)	4.0×10^{-10}	4.8×10^{-10}	3.7×10^{-10}
$T_{1/2}$(–EGF)(h)	21 ± 4	20 ± 5	15 ± 2
B:			
TGFα mRNA	++	+	+
TGFβ mRNA	++	+	+

of growth control. First, the EGFR in subclone S4 might have a subtle alteration in affinity or half-life, causing an increased requirement for EGF; or second, subclone S4 might depend less on the production of autocrine growth factors than the other clones.

As shown from the data in Table 2, neither of these possibilities seems to be operating in this case. The receptors in S4 and S5 are indistinguishable in terms of frequency, affinity, and half-life. In addition, the receptors of both clones do not differ from the parental cells in terms of affinity and half-life. We have also quantitated steady state levels of transcripts of the TGFα and TGFβ genes; the secreted growth factor products of these genes have been implicated in the growth control of breast cancer cells in tissue culture (21), including MDA-MB-468 (22). Although expression of both genes in the subclones is less than in the parental cells, no differences between S4 and S5 could be detected. This, of course, does not rule out differences at the protein level of these potentially autocrine growth regulators since regulation of their production might be posttranscriptional or posttranslational under some circumstances.

On the basis of these preliminary experiments, we believe that the mechanism involved in the EGF-dependency of subclone S4 likely lies in other gene expression that is regulated by EGF specifically in anchorage-independent tissue culture. Clone S4 may therefore represent a useful source of information on gene expression necessary for anchorage-independent growth.

MECHANISMS OF EGF SIGNAL TRANSDUCTION IN MDA-MB-468

The intracellular events contributing to signal transduction from the EGFR are not totally understood. A number of events subsequent to EGF binding have been implicated, including phosphorylation of a number of protein substrates, receptor autophosphorylation, activation of the Na^+/H^+ exchange system resulting in cytoplasmic alkalinization, mobilization of intracellular calcium stores, stimulation of phosphatidyl inositol (PI) turnover, and changes in gene expression (particularly induction of expression of c-*myc* and c-*fos* genes) (1). Of these events, only activation of the receptor kinase has been shown to be essential for signal transduction since various mutant EGFRs devoid of kinase activity are unable to transduce signals (23). PI turnover is frequently activated by EGF stimulation, particularly in cells overexpressing EGFR (24). An important series of experiments has suggested that in A431 cells the linkage between EGFR and the generation of second messengers may be achieved through the fact that phospholipase C (PLC), the enzyme responsible for PI turnover, is a substrate for the tyrosine phosphorylation catalyzed by the activated receptor (25). We have confirmed this finding in MDA-MB-468 breast cancer cells by immunoprecipitating a tyrosine-phosphorylated form of PLC subsequent to EGF binding (Church, Pawson, and Buick, unpublished observations). The functional significance of tyrosine phosphorylation of PLC remains to be established.

The growth properties of MDA-MB-468 and subclones provided an opportunity to assess differential mechanisms of signal transduction under circumstances of negative and positive growth regulation. For example, in these cell lines, as in other cells responsive to EGF, an early consequence of EGFR activation is the transient accumulation of transcripts of the c-*myc* and c-*fos* genes. It is considered possible that the products for these genes act to regulate transcription of other genes and, thus, play a role in the altered gene expression elicited by EGF. However, we found that the kinetics of the altered accumulation of c-*myc* and c-*fos* transcript was identical for all cell lines: maximum at 30' for c-*myc* mRNA and 15' for c-*fos* mRNA. In addition, we found that the magnitude of the increased accumulation was similar in the parental cell line and subclones, despite the fact that EGF is acting as growth inhibitor and stimulator, respectively (13). Therefore, the transient elevation in level of transcript and subsequent protein expression from the c-*myc* and c-*fos* genes does not seem to play a role in the selectivity of the action of EGF in MDA-MB-468 and subclone S4.

Further evidence that dissociates c-*myc* and c-*fos* transcription from EGF-induced growth inhibition or mitogenesis came from experiments designed to assess the sensitivity of MDA-MB-468 and subclone S4 to pertussis toxin (PT). We initiated these experiments since many hormone receptor systems are linked to the generation of second messengers by the interaction of G-proteins. This family of proteins has the property of linking receptor occupancy to activation of adenyl cyclase, PLC, or ion channels (26). A feature of certain G-protein α-subunits is a sensitivity to ADP-ribosylation by (PT); demonstration of an attenuation of hormone effect by PT has therefore been used to define the role of G-protein intermediates in the signal transduction of various hormones (26).

We demonstrated that PT could block both EGF-induced growth inhibition in MDA-MB-468 in anchorage-dependent or independent culture and the EGF-dependent growth of clone S4 in anchorage-independent conditions (27). This indicated the possibility of a G-protein intermediate in EGF signal transduction. A role for such intermediates has also been suggested by the demonstration of EGF-mediated phosphorylation of a G-protein β-subunit in human placental membranes (28) and by the PT sensitivity of proliferative responses of hepatocytes to EGF (29). The molecular characteristics of the G-protein intermediate in MDA-MB-468 cells have not been elucidated. Under the conditions of the experiments, PT catalyzed the transfer of ADP-ribose from NAD^+ to an intrinsic membrane protein(s) of approximately 40 Kd. Therefore, it is possible that the target for the ADP-ribosylation of the PT is an α-subunit of one of the trimeric G-protein family. Since cAMP is not affected under these conditions, the effective G-protein is not the G_i-protein coupling to adenyl cyclase. It is of interest that the G-protein-mediated pathway of EGF signal transduction may have some specificity for this particular cell line since we have shown that EGF growth inhibition of A431 cells is not PT sensitive.

Although PT is able to block the proliferative changes associated with EGF exposure in MDA-MB-468 and subclone S4, these same conditions did not block the increased accumulation of c-*myc* and c-*fos* transcripts (27). This implies either (*a*) that the PT-sensitive event exists on a separate pathway of transduction from that causing the transcription of c-*myc* and c-*fos* genes, or (*b*) that the PT-sensitive event exists on the same pathway as that causing c-*myc* and c-*fos* transcription, but at a distal position. Given the traditional role of G-protein intermediates in coupling activated membrane receptors to second-messenger systems, the multiple-pathway model seems to be the most probable. We are currently working to assign early biochemical events in EGF signal transduction to PT-sensitive or -insensitive classes of transducing pathways.

One early consequence of EGF binding is cytoplasmic alkalinization caused by the activation of the Na^+/H^+-exchange antiport. Such early changes also occur in both MDA-MB-468 and subclone S4 in response to EGF despite the opposite proliferative response of the two cell lines (30). We have used two lines of evidence to rule out an obligatory role for such activation in the EGF signal transduction process.

First, blocking Na^+/H^+ exchange in both cell lines with amiloride analogues or by manipulating extracellular pH failed to block the ability of EGF to inhibit growth of MDA-MB-468 or stimulate growth of subclone S4. Second, activation of Na^+/H^+ exchange is unable to alter accumulation of c-*myc* and c-*fos* transcripts in the absence of EGF. The Na^+/H^+-exchange system therefore seems to be activated by a separate EGF-mediated transduction system distinct from those responsible for growth inhibition/stimulation or for c-*myc*/c-*fos* transcription. A summary of the characteristics of EGF signal transduction in MDA-MB-468 and subclone S4 is shown in Table 3.

CONCLUSION

The breast cancer cell line MDA-MB-468 has provided a useful model of the abnormal growth regulation elicited by overexpression of the EGFR. It is possible that a high proportion of estrogen-receptor negative breast tumors have acquired an in vivo growth advantage through this mechanism. The characteristics of MDA-MB-468 and variant subclones have allowed for demonstration of a relationship between tumor growth rate in vivo and level of EGFR expression. In addition, the cell lines have been useful for a dissection of the biochemical events leading to EGF signal transduction. In particular, these cell lines have a reliance on a novel G-protein-dependent pathway to couple the activated EGFR to signal processing.

Table 3. Characteristics of EGF signal transduction in MDA-MB-468 and subclone S4.

A: Accumulation of transcripts of c-myc *and* c-fos *genes*

1. It occurs in MDA-468 and MDA-468-S4 with identical kinetics after EGF binding (max. 15' for c-*fos* and 30' for c-*myc*).
2. It is not affected by pertussis toxin under conditions that demonstrate proliferative effects of EGF.
3. It is not induced by activation of Na^+/H^+ in the absence of EGF.

B: Activation of Na^+/H^+ *exchange*

1. Amiloride analogues (EPPA and EPA) block EGF-induced cytoplasmic alkalinization, but do not affect EGF-induced growth inhibition of MDA-468 or growth stimulation of MDA-468.
2. Low-pH media abrogate EGF-induced cytoplasmic alkalinization, but do not affect growth inhibition or stimulation.
3. Activation of Na^+/H^+ exchange is not able to induce c-*myc* and c-*fos* transcription.

REFERENCES

1. Carpenter G. Receptors for epidermal growth factor receptor and other polypeptide hormones. Annu Rev Biochem 1987;56:881-914.
2. Downward J, Yarden Y, Mayes E, et al. Close similarity of EGF receptor and V-erb-B oncogene protein sequences. Nature 1984;307:521-7.
3. Okamoto S, Oka T. Evidence for physiological function of EGF: Pregestational sialoadenectomy of mice decreases milk production and increases offspring mortality during lactation period. Proc Natl Acad Sci USA 1984;81:6059-63.
4. Kurachi H, Okamoto S, Oka T. Evidence for involvement of the submandibular gland EGF in mouse mammary tumorigenesis. Proc Natl Acad Sci USA 1985;82:5940-3.
5. Vonderhaar BK, Nakhasi HL. Bifunctional activity of EGF on α and K-casein gene expression in rodent mammary glands *in vitro*. Endocrinology 1986;119:1178-84.
6. Stampfer, 1990 (this volume).
7. Fitzpatrick SL, LaChance MP, Schultz GS. Characterization of EGF receptor and action on human breast cancer cell in culture. Cancer Res 1984;44:3442-7.
8. Sainsbury JRC, Farndon JR, Sherbet GV, Harris AL. EGF receptors and estrogen receptors in human breast cancer. Lancet 1985;1:364-6.
9. Ozanne B, Richards CS, Hendler F, Burns D, Gusterson B. Over-expression of the EGF receptor is a hallmark of squamous cell carcinomas. J Pathol 1986;149:9-14.
10. Liberman TA, Nusbaum HR, Razon N, et al. Amplification, enhanced expression and possible re-arrangement of the EGF receptor gene in primary human brain tumors of glial origin. Nature 1985;313:144-7.
11. Harris AL, Nicholson S. Epidermal growth factor receptors in human breast cancer. In: Lippman, Dickson, eds. Breast cancer: Cellular and molecular biology. Boston: Kluwer Academic, 1988:93-118.

12. Filmus J, Pollak MN, Cailleau R, Buick RN. MDA-468, a human breast cancer cell line with a high number of EGF receptors, has an amplified EGF receptor gene and is growth inhibited by EGF. Biochem Biophys Res Comm 1985;128:898-905.
13. Filmus J, Trent JM, Pollak MN, Buick RN. The EGF-receptor gene-amplified MDA-468 breast cancer cell line and its non-amplified variants. Mol Cell Biol 1987;7:251-7.
14. Kudlow JE, Cheung C-YM, Bjorge JD. EGF stimulates the synthesis of its own receptor in a human breast cancer cell line. J Biol Chem 1986;261:4134-8.
15. Kawamoto T, Mendelsohn J, Le A, Sato GH, Lazar CS, Gill GN. Relation of EGF receptor concentration to growth of human epidermoid carcinoma A431 cells. J Biol Chem 1984;259:7761-6.
16. Kamata N, Chida K, Rikimaru K, Horikoshi M, Enomoto S, Kuroki T. Growth inhibitory effects of EGF and overexpression of its receptors on human squamous cell carcinomas in culture. Cancer Res 1986;46:1648-53.
17. Haley JD, Hsuan JJ, Waterfield MD. Analysis of mammalian fibroblast transformation by normal and mutated human EGF receptors. Oncogene 1989;4:273-83.
18. De Fiore PP, Pierce JH, Fleming TP, et al. Overexpression of the human EGF receptor confers an EGF-dependent transformed phenotype to NIH 3T3 cells. Cell 1987;51:1063-70.
19. Santon JB, Cronin MT, Macleod CL, Mendelsohn J, Masui H, Gill GN. Effects of EGF receptor concentration on tumorigenicity of A431 cells in nude mice. Cancer Res 1986;46:4701-5.
20. Lee E Y-HP, To H, Shew J-Y, Bookstein R, Scully P, Lee W-H. Inactivation of the retinoblastoma susceptibility gene in human breast cancers. Science 1988;241:218-21.
21. Dickson RB, Lippman ME. Control of human breast cancer by estrogen, growth factors and oncogenes. In: Lippman ME, Dickson RB, eds. Breast cancer: Cellular and molecular biology. Boston: Kluwer Academic, 1988:119-65.
22. Fernandez-Pol J-A, Klos DJ, Hamilton PD, Talkad VD. Modulation of EGF receptor gene expression by TGFβ in a human breast carcinoma cell line. Cancer Res 1987;47:4260-5.
23. Honegger AM, Szapary D, Schmidt A, et al. A mutant EGF receptor with defective protein tyrosine kinase is unable to stimulate proto-oncogene expression and DNA synthesis. Mol Cell Biol 1987;7:4568-71.
24. Wahl MI, Sweatt JD, Carpenter G. EGF stimulates inositol triphosphate formation in cells which overexpress the EGF receptor. Biochem Biophys Res Comm 1987;142:688-95.
25. Wahl MI, Nishibe S, Suh P-G, Rhee SG, Carpenter G. EGF stimulates tyrosine phosphorylation of phospholipase c-II independently of receptor internalization and extracellular calcium. Proc Natl Acad Sci USA 1989;86:1568-72.
26. Milligan G. Techniques used in the indentification and analysis of function of pertussis-toxin-sensitive guanine nucleotide binding proteins. Biochem J 1988;255:1-13.
27. Church J, Buick RN. G-protein mediated EGF signal transduction in a human breast cancer cell line. J Biol Chem 1988;263:4242-6.
28. Valentine-Braun KA, Northup JK, Hollenberg MD. EGF-mediated phosphorylation of a 35 kda substrate in human placental membranes; relationship to the β subunit of the guanine nucleotide regulatory complex. Proc Natl Acad Sci USA 1986;83:236-40.

29. Johnson RM, Connelly PA, Sisk RB, Pobiner BF, Hewlett EL, Garrison JC. Pertussis-toxin or phorbol 12-myrtistate 13-acetate can distinguish between EGF and angiotensin-stimulated signals in hepatocytes. Proc Natl Acad Sci USA 1986;83:2032-6.
30. Church J, Mills GB, Buick RN. Activation of the Na^+/H^+ antiport is not required for EGF-dependent gene expression, growth inhibition or proliferation in human breast cancer cells. Biochem J 1989;256:151-7.

9

Mammary Growth Regulation by Transforming Growth Factor β

Charles W. Daniel and Gary B. Silberstein

Department of Biology, Sinsheimer Laboratories, University of California, Santa Cruz

A considerable literature has developed around the problem of regulation of mammary development. The great majority of this work has concerned the identification of systemic mammogens, notably the ovarian steroids, pituitary peptides, adrenal corticoids, and placental factors (1). The distinguishing feature of these hormones is that they all drive mammary development towards growth or functional differentiation; that is, they act as positive regulators. The existence of these hormones has been, by and large, demonstrated in vivo by endocrine ablation surgery followed by replacement therapy, in which removal of endocrine secretion interrupted the normal development or function of the organ. Because these experiments were conducted in a physiological setting, their biological significance was never in doubt. The present situation is that the naturally occurring mammogens have been identified and described in considerable detail and, when more recent biochemical and molecular approaches are factored in, it can be said that we are developing a usable outline of the hormones driving mammary growth and function.

A much less satisfactory situation prevails when one asks about the other side of the developmental coin, the negative regulation of growth and morphogenesis. In this, the mammary gland is similar to most other tissues and organs—we tend to know much more about factors that stimulate development than those that limit it. With respect to mammary development in the subadult and adult mouse, with which this paper is largely concerned, four statements concerning the nature of growth regulation in the mammary gland can be made on the basis of the earlier literature.

First, the growth of mammary ducts in the virgin animal is not limited by an intrinsic inability of the tissue to proliferate further as, for example, might be the

---

Acknowledgments: This work was supported by PHS grant CA-45231 from the National Cancer Institute.

case in a tissue that undergoes terminal differentiation following a programmed period of growth. The proliferative potential of mature mammary tissue cells has been demonstrated in an important series of experiments (2–3) in which fragments of growth-quiescent mammary ducts were transplanted into mammary gland-free fat pads in which the embryonic anlage of the mammary tree had been surgically removed, leaving vacant mammary fat pad available as substrate. Mammary implants rapidly regenerate and grow to fill the available area of the "cleared" fat pads. Indeed, mammary tissue is able to fill a series of cleared fat pads during serial transplantation before senescent changes eventually limit growth (4). These experiments demonstrate that the proliferative potential of mammary gland cells is many times that required during the life span of a mouse. They also demonstrate that stem cells capable of regenerating the entire, functionally competent gland are found throughout the ductal tree.

Second, negative regulation does not operate through systemically distributed hormones. During early development primitive mammary ducts grow into the mammary fat pad, aggressively invading the adipose stroma and forming an arborizing mammary tree (5). At approximately 8 weeks of age, this process is terminated when the gland reaches the limits of the mammary fat pads. Using the transplant techniques previously described, samples of gland from such an animal can be transplanted into a cleared fat pad in the same animal. Growth and morphogenesis recommences in the transplanted gland, while gland in the fully occupied fat pad remains quiescent. One concludes that limitations upon mammary growth must be of local origin, and represents a process of glandular autoregulation.

Third, the two processes of autoregulation and pattern formation in the mammary ductal tree are related and inseparable. Mammary transplants are unable to grow when surrounded by preexisting mammary tissues, and mammary buds exhibit avoidance behavior, turning to avoid mammary structures (2). These inhibitory and turning behaviors can be observed in the same gland in which, a few millimeters away, mammary ducts are actively elongating into unoccupied fatty stroma. It is apparent that regulation involves the action of local mechanisms that operate over a distance of 1 or 2 mm and that these mechanisms result in the production of an open pattern of arborization in which secretory alveoli may later develop. The maintenance of pattern in the mammary gland is as important as its formation, and pattern maintenance is clearly an active, rather than a passive, process; this is apparent from the previously mentioned regeneration experiments. In normal gland, it seems, ducts inhibit lateral budding of neighboring ducts, and the adipose tissue between them is rendered nonpermissive for growth.

Fourth, mammary regulation probably operates through diffusible autocrine or paracrine factors. From arguments given above, it can be inferred that mammary regulation is a short-range, local phenomenon. The possibility that regulation could occur by cell-cell communication through specialized membrane junctions has been investigated by electron microscopy. Although epithelial cells within ducts are tightly associated and display typical communicating junctions (6), there appears to be no direct cellular contact between ducts. The basal lamina surrounding ducts and end buds is intact, and no cytoplasmic processes penetrating it have been observed (7), suggesting that interductal communication probably occurs by means of diffusible factors.

METHODS: SLOW-RELEASE IMPLANTS

Possible effects of growth factors are usually tested by injection. Results so obtained are difficult to interpret and have been of limited usefulness. As a consequence, in vitro techniques have provided most of our knowledge concerning the effects of growth factors on tissues and cells.

A method for the sustained, localized delivery of growth factors would appear to be a desirable alternative, in which the direct effects of these molecules can be tested in a physiological setting. The most fully investigated delivery system is the slow-release plastic EVAc (ethylene vinyl acetate copolymer; Dupont Chemical Co., Universal City, CA). This material is capable of releasing substances in unaltered, bioactive form. EVAc alone is remarkably inert, stimulating no inflammatory response when implanted into tissues and demonstrating no detectable intrinsic bioactivity. EVAc can be placed directly into the tissue in the area of interest, where it gradually releases its contents for the duration of the experiment. The release of organic-soluble materials, such as steroids, is very gradual (8), whereas water-soluble materials, such as peptides, display a biphasic pattern, with an immediate surge of released material followed by a slower, prolonged release period (9–10). Another aspect of local delivery systems requires comment. When materials are injected they may act briefly in a local fashion at the site of injection, but they quickly enter the vascular compartment, and their principle effect is systemic. Any movement of material into the interstitial compartment where interactions with tissue cells are possible is mediated by vascular permeability. In contrast, material released from EVAc enters the interstitial spaces directly, where interactions with surrounding cells and tissues can occur; some passage into the vascular system is found, of course, but at low doses this rarely interferes with the ability to interpret a response as direct. That is, if the tissue around the implant is responsive, but tissue in another gland in the same animal is not, the response can safely be considered as local and direct.

We have used EVAc slow-release implants to examine the direct effects of a number of growth factors, pharmacological agents, and hormones (11–12) (Fig. 1). Most agents either stimulated the gland or had no effect. Among those showing growth inhibition, epidermal growth factor and transforming growth factors β_1 and β_2 appeared to be of particular interest. In this chapter our progress in understanding the role of TGFβ_1 is reviewed.

EFFECTS OF EXOGENOUS TGFβ_1

TGFβ_1 is a member of a family of dimeric peptides now numbering five closely related, but clearly distinct molecular forms. These five molecular forms belong, in turn, to a large supergene family of peptides identified in a variety of species, all of which share the property of being apparently concerned with processes of develop-

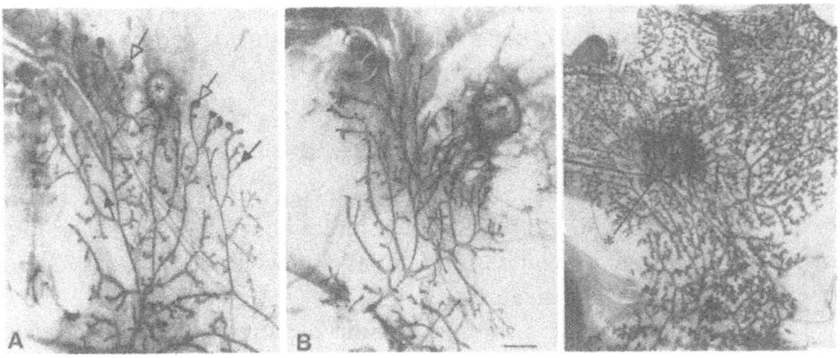

Fig. 1. Effects of TGFβ_1 on ductal and lobulo-alveolar growth. *A*: Control EVAc pellet (*) containing BSA. Open arrows indicate rapidly growing end buds. *B*: EVAc (*) containing 150-ng TGFβ_1 for 2 days. End buds have become simple ductal tips. *C*: EVAc pellet containing TGFβ_1 has no effect on alveolar development. (Bar = 1 mm.) (From Daniel CW, Silberstein GB, Van Horn K, Strickland P, Robinson S, TGF-beta-1 induced inhibition of mouse mammary ductal growth: Developmental specificity and characterization, Dev Biol 1989; 135:20–30, with permission.)

mental regulation. Knowledge of the TGFβ group has undergone an explosive increase in recent years and now constitutes a large field of investigation (13) that we shall not attempt to review beyond pointing out two significant generalizations. First, although TGFβ_1 was originally identified as a mitogen for cultured fibroblasts, it has subsequently been shown to be growth inhibitory for most epithelial cells in culture. Its suppression of growth of mouse mammary epithelium is not surprising in view of its antimitotic effects on cultured mammary cells (13).

Second, it is useful to point out that the TGFβ family of growth factors was discovered in vitro, and most of our knowledge concerning its action has been derived from in vitro studies. In this it resembles other growth factors, and like others, there is considerable interest in identifying its biological roles. Identification of TGFβ_1 as a mammary regulator would therefore be of considerable interest to specialists in the TGFβ field, as well as to those who are mainly concerned with mammary gland biology.

Characteristics of Inhibited Tissues

The effect of TGFβ_1 on ductal growth was investigated by implanting small pieces of EVAc containing the growth factor in the path of the advancing gland, directly in front of the mammary end buds (Fig. 1). The end buds, which represent the growth points of the gland, when treated with TGFβ_1, rapidly regressed into blunt-ended terminal structures that were similar in appearance to terminal ducts in other regions of the gland where end bud growth had been inhibited by normal growth regulatory processes (Fig. 1, open arrows).

Table 1. Summary of growth effects of implanted hormones and factors.

Agent	Effect		
	End Buds[a]	Lobules[b]	Reference[b]
Classical mammogens			
17β-estradiol	+	0	8
17α-estradiol	0	0	8
Progesterone	0	0	
DCA	+	0	14
Growth hormone	+	0	11
Prolactin	+	0	11
Nontraditional mammogens			
Cyclic AMP agents	+	0	15
EGF	+	+	9, 16
TGFα	n/a	+	16
Nonmammogenic agents			
Insulin	0	0	
Basic FGF	0	0	
PDGF	0	0	
$TGF\beta_1$ + EGF in ovx	0	n/a	17
$TGF\beta_1$ – EGF in 5 weeks	–	0	10, 17
EGF in ovx	+	n/a	9
EGF in 5 weeks	–	n/a	18
Glucocorticoids	–	n/a	
Testosterone	0	n/a	
BSA control	0	0	11

[a] To test for stimulation, ovariectomized animals in which the ductal system had completely involuted were implanted on one side only, with the contralateral gland serving as a control.

[b] To test for lobulo-alveolar differentiation, implants were placed in 5-week or 3-month-old hormonally intact animals.

Key: + = stimulation; – = inhibition; 0 = no effect; n/a = not applicable.

Note: Results for agents not referenced in the table are unpublished observations of the authors.

Ductal growth inhibition is a normal feature of glandular development, and by several criteria, the $TGF\beta_1$-treated ducts and ductal tips resembled the growth-quiescent ducts in untreated glands. Both treated and untreated terminal ducts were blunt-tipped, thin-walled structures surrounded by a thin layer of fibrous connective tissue. There was no apparent tissue disruption or necrosis, as might be predicted if $TGF\beta_1$ produced a cytotoxic effect. We concluded that at the time points examined, 2 and 4 days, the inhibited structures were apparently normal. As will be seen in a subsequent section, earlier time points revealed unusual fibrous condensations in the stroma.

Growth Inhibition Is Reversible.

Growth-quiescent ductal tissue is not terminally differentiated, but appears to be inhibited from further growth. This inhibition must be an active, fully reversible process, in which inhibition is a consequence of the density of ductal elements in the mammary fat pad. This is apparent from previously mentioned transplantation experiments, and it is also apparent from consideration of mammary development during early pregnancy when, as a result of endocrine changes, the gland undergoes a burst of ductal growth (fine branching) followed by alveolar morphogenesis and functional differentiation.

Reversibility of the TGFβ_1 effect was demonstrated by removing the implants from inhibited glands and showing that normal growth had resumed (10, 17) (Fig. 2), indicating that TGFβ_1 did not lead to the terminal differentiation observed with certain cultured cell lines (13).

Does the Tissue Become Refractory?

If TGFβ_1 is to be considered a likely candidate for homeostatic tissue regulation, active growth inhibition must be effective for extended periods—the life of the animal, in nonbreeding females. In order to determine whether mammary ductal tissue became refractory to extended exposure, two experiments were carried out (Fig. 2). In the first experiment, a second TGFβ_1 treatment immediately followed recovery from a first, and we found that ductal growth was again inhibited, followed

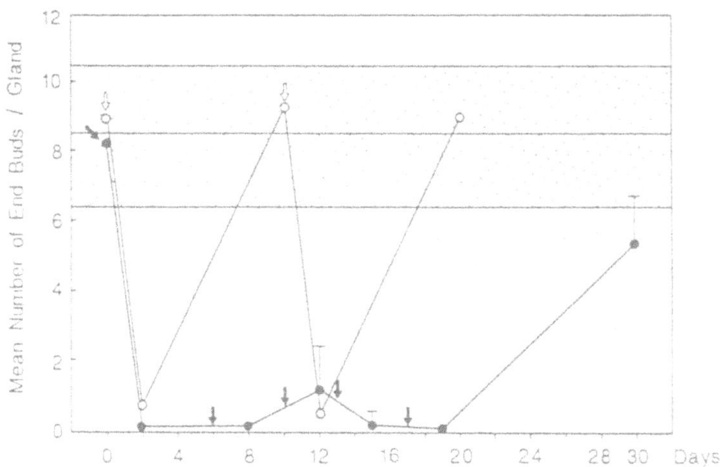

Fig. 2. Effects of multiple treatments. (Open circles = treatment recovery cycle; closed circles = repeated, prolonged exposure; shaded area = 95% confidence interval for control values.) (From Daniel CW, Silberstein GB, Van Horn K, Strickland P, Robinson S, TGF-beta-1 induced inhibition of mouse mammary ductal growth: Developmental specificity and characterization, Dev Biol 1989; 135:20–30, with permission.)

in a few days by another recovery. This suggested that the gland is capable of repeated cycles of inhibition and growth. In a second test, glands treated at frequent intervals for 20 days remained inhibited for the entire period of exposure, after which they recovered to control values. In neither case was there an indication of diminished sensitivity to TGFβ_1. This capacity for sustained tissue responsiveness to exogenous TGFβ_1 contrasts with other peptide growth factors whose transient effects are due to receptor down-regulation, a phenomenon we have recently demonstrated with EGF in the mammary gland (18).

Time-Course and Dose-Response of TGFβ$_1$ Inhibition

Endogenous local regulators must be able to exert their inhibitory effects rapidly enough to account for the observed patterning of the ductal tree. Because of the extremely rapid growth rate of large end buds, inhibition should probably take effect in the 12- to 24-h time range. Our time course studies indicated that end bud numbers were reduced about 50% by 24 h and fully inhibited by 2 days (10). When we examined epithelial DNA synthesis, a more rapid response was evident, in which the DNA labeling index of end bud cells was reduced to low values by 12 h. The inhibitory effect appears to occur rapidly enough to generate the observed pattern of mammary ducts, in which growth stops short of the fat pad limits, and ductal crossing is only occasionally observed.

Using both EVAc implants and direct injection into the mammary fat pad, inhibition was found to occur over a very narrow concentration range of 25–50 ng/pellet, or 0–9 ng/injection (10). Injections of 5 ng were capable of extensive growth inhibition. With respect to actual amounts to which responding cells are exposed, the complexity of the system makes rigorous estimates impossible. Nevertheless, it is interesting to consider that injected growth factor must diffuse throughout the gland, becoming trapped, diluted, and washed out by capillary action in the process. It is reasonable to conclude that several millimeters away from the site of a 5-ng injection, the local interstitial concentrations are small, probably comparable with the 1- to 5-ng/mL concentrations that have been shown to inhibit a variety of cells in tissue culture (19). We conclude, first, that the steep dose-response curve (10), almost a trigger effect, is fully compatible with the action of a hypothetical regulator and, second, the minimum effective dosages are estimated to be small, comparable with observed tissue culture dosages.

Specificity of the Inhibitory Response

Is the TGFβ_1 effect found in other stages of the mammary cycle? To date we have investigated this question only in alveolar tissues, generated either by pregnancy or by systemic treatment with estradiol and progesterone. The results were clear-cut, indicating a lack of effect on alveolar development at dosages that were fully inhibitory to ductal tissue (10) (Fig. 1).

In related experiments we investigated the effects of TGFβ_1 on growth and DNA synthesis in mature ducts, in contrast to earlier experiments focusing on end buds. The question of interest was whether exogenous TGFβ_1 is capable of suppressing lateral budding along the ductal tree, and if so, did it suppress all

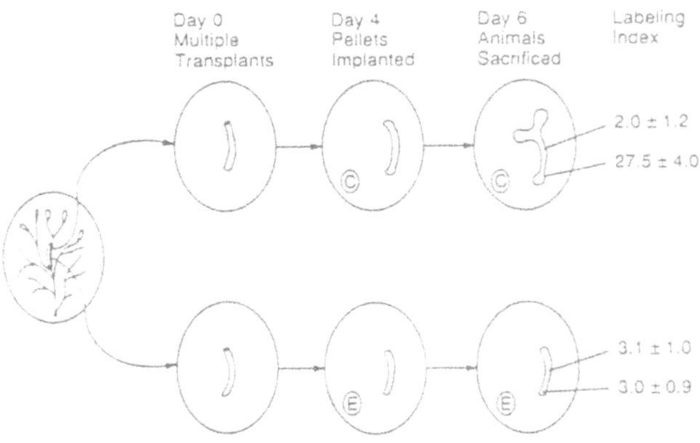

Fig. 3. Effects of TGFβ_1 on growth and DNA synthesis in mature ducts. Transplants were made on day 0, treated for 2 days, and terminated on day 6. (C = control; E = TGFβ_1; labeling index is a measure of the replicating pool.)

epithelial cell division or only that connected with growth of lateral buds and branches? This question was tested in a transplant experiment (Fig. 3), in which ductal segments were placed into cleared fat pads, permitting formation of growth buds. The conclusion from this study is that TGFβ_1 completely suppresses the formation of new buds. "Maintenance" DNA synthesis, on the other hand, which is concerned with normal cell turnover, is unaffected. Thus, TGFβ_1 displays remarkable specificity and appears to be selective for stem cells. This conclusion is reinforced by the observation that in TGFβ_1-treated end buds, the cells that appear to be affected earliest and most obviously are the undifferentiated cap cells, identified as myoepithelium progenitor cells (7).

ENDOGENOUS TGFβ_1

TGFβ_1 Gene Transcription in Mammary Tissues

Inhibition of mammary development by TGFβ_1 does not in itself, of course, indicate a physiological role for the growth factor. We have therefore examined mammary tissues at various developmental stages for TGFβ_1 gene transcription and for the presence of immunoreactive growth factor.

Northern hybridization was carried out on blots of total glandular RNA, using a ^{32}P RNA probe containing a 278-base sequence coding for a region of the active form of murine TGFβ_1. Results indicate that substantial amounts of TGFβ_1 mRNA are found in the mammary gland, more than in the HSV 3T3 cells used as a positive control. The distribution of mRNA in glands at various stages of development shows the largest amount in actively growing, virgin gland, and in gland from ovariectomized mice (Fig. 4). These results are consistent with the interpretation that TGFβ_1 may play a regulatory role.

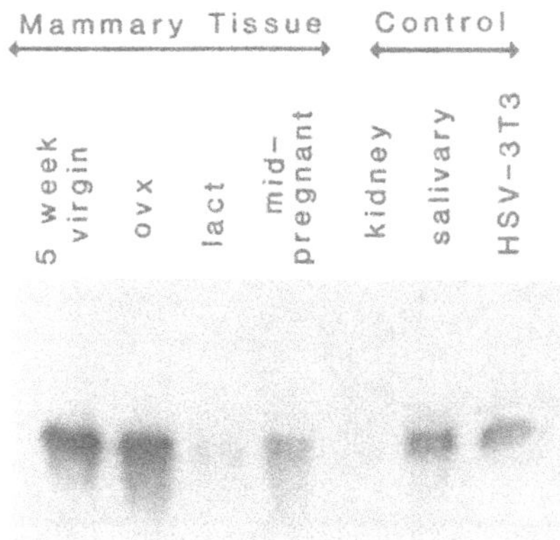

Fig. 4. Northern blot of total RNA from glands in different growth states and controls. (Lane 1 = 5-week-old virgin [end buds]; lane 2 = ovariectomized; lane 3 = lactating; lane 4 = midpregnant; lane 5 = kidney; lane 6 = salivary gland; and lane 7 = HSV 3T3 cells.) RNA of 10 μg was loaded per lane, and loading variation was checked with ethidium bromide stain (not shown).

Hybridization in situ to glandular $TGF\beta_1$ mRNA has been carried out using an RNA probe containing a sequence coding for a portion of the active form of the molecule. Positive controls consisted of embryonic cartilage-forming tissues (20) and adult skin induced or not induced with phorbol esters (21). Viral gene sequences served as negative controls. Hybridizations resulted in substantial labeling in the epithelial end buds and lighter but still significant labeling in ductal epithelium and stromal cells (Fig. 5). Particularly interesting is the presence of "hot spots" in the end buds, which we hope to be able to correlate with observed

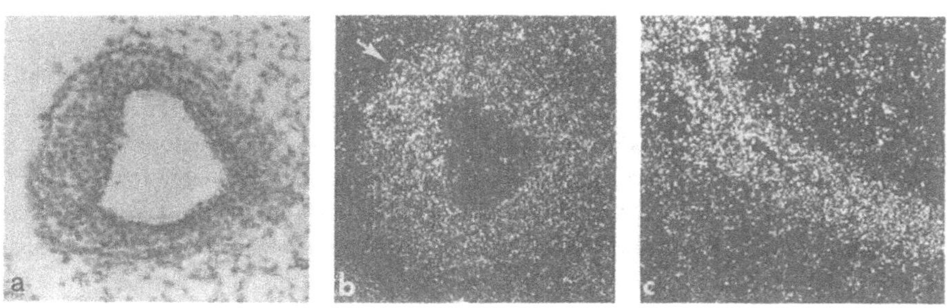

Fig. 5. Sections of mammary end buds and ducts hybridized with a ^{35}S-labeled RNA probe to the coding region of mature $TGF\beta_1$ (175×). Epithelium of end buds and ducts are labeled, and some "hot spots" are evident. Several positive and negative controls have been used (not shown).

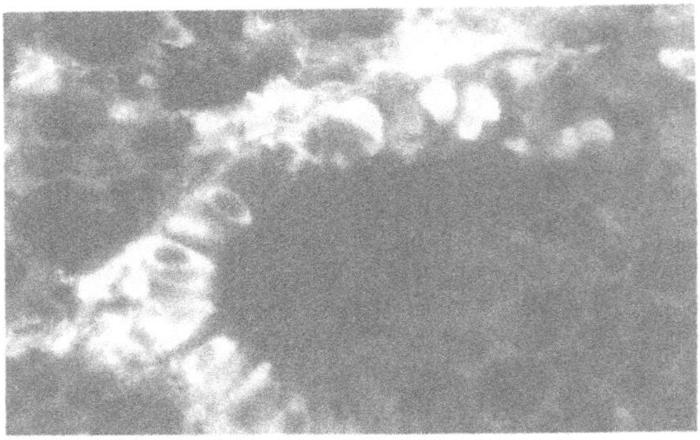

Fig. 6. Immunolocalization of an epitope to the mature domain TGFβ_1.

morphogenetic events, such as ductal bifurcation or turning. In any case, the presence of abundant TGFβ_1 mRNA in areas of morphogenetic activity has been confirmed.

Immunolocalization of TGFβ_1

TGFβ_1 is synthesized as a precursor that is latent, or inactive; cleavage of this precursor by proteases or acid produces the active factor, also referred to as mature or processed. Using a battery of antibodies to synthetic peptides representing epitopes in the precursor and mature domains, the distribution and sites of synthesis of TGFβ_1 in the gland are beginning to emerge (Fig. 6). Using the peroxidase-diaminobenzidine reaction to visualize the immune complex, we found staining for the precursor molecule to be localized mainly intracellularly, with the largest amounts in the end bud epithelium. The mature epitope was localized in the periductal stroma and also in epithelium of ducts and end buds. The distribution of the mature epitope was not uniform, with small clusters of heavily stained epithelial cells distributed along the ducts. Significantly, little or no immunoreactive material was found in the stroma at the end bud tip.

Taking also into account the molecular hybridization data, these results suggest that epithelial cells of the end buds, and to a lesser extent those of the ducts, synthesize and secrete TGFβ_1 that is then complexed to matrix molecules, such as fibronectin (22), or to integrin through RGD sequences in the latent form of TGFβ_1 (13). We hypothesize that this immobilized TGFβ_1 in the periductal stroma functions to suppress lateral adventitious budding, which would infill the interductal spaces and destroy the open growth pattern of the ductal tree (Fig. 7).

Induction of Mammary Extracellular Matrix by TGFβ_1

TGFβ_1 stimulates the synthesis and deposition of extracellular matrix (ECM) both in vitro and in vivo (13) and hastens wound healing by promoting the synthesis of fibrous connective tissues (23). This knowledge led us to examine the early stages of

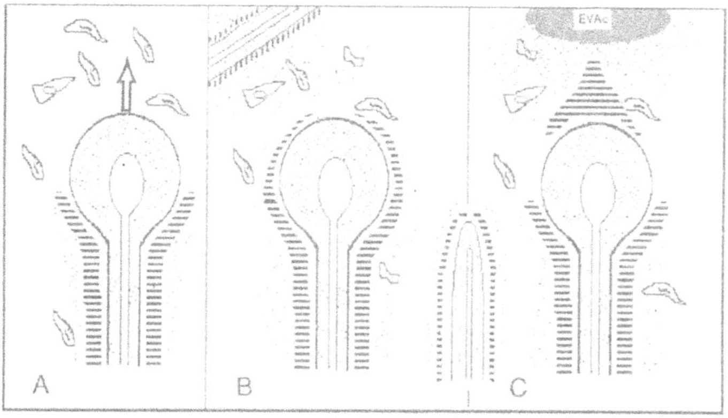

Fig. 7. Proposed model of mammary regulation by TGFβ_1 (dots). *a*: Rapidly growing end bud secretes TGFβ_1 along its sides and around ducts, stimulating formation of growth-suppressive ECM. *b*: Approaching a competing duct, the end bud reacts to the presence of TGFβ_1 from the competing duct by reducing growth. Fibrous ECM advances along the sides of the end bud, eventually surrounding it. *c*: EVAc implant releases large amounts of TGFβ_1 in an area where it is not ordinarily found, at the end bud tip, causing formation of a fibrous cap. End buds in both (*b*) and (*c*) later regress to form terminal structures encased in fibrous matrix, similar to that shown between the two structures.

mammary growth inhibition by TGFβ_1 for possible effects on mammary ECM that might shed light on mechanisms of growth regulation.

As background to this study, we point out that the normal branching morphogenesis of mammary ducts, as well as the normal process of growth inhibition, involves a complex interplay of epithelium and stroma. This is manifested by the replacement of fatty connective tissue by fibrous ECM at the interface of the two tissues (7, 24–25). It is also clear that the elaboration of mammary-specific ECM is an example of an inductive interaction. As is the case with other examples of epithelial-mesenchymal inductive interactions, the molecular signals responsible for regulating the synthesis, localization, and deposition of ECM are unknown. This fibrous condensate forms a continuous sheath along the entire ductal tree, the only exceptions being the tips of growing end buds and occasional lateral buds, where the basal lamina of the epithelium is in direct contact with adipocytes. Thus, there is a direct correspondence between the presence of a fibrous tunic, growth suppression, and immunoreactive TGFβ_1.

Our studies of the early stages in TGFβ_1 ductal inhibition led to the finding that within the time period around 24 h, mammary end buds within the zone of influence of implanted TGFβ_1 displayed at their tips a zone of activated stroma that was not present, or was present in limited degree, in untreated gland. This took the

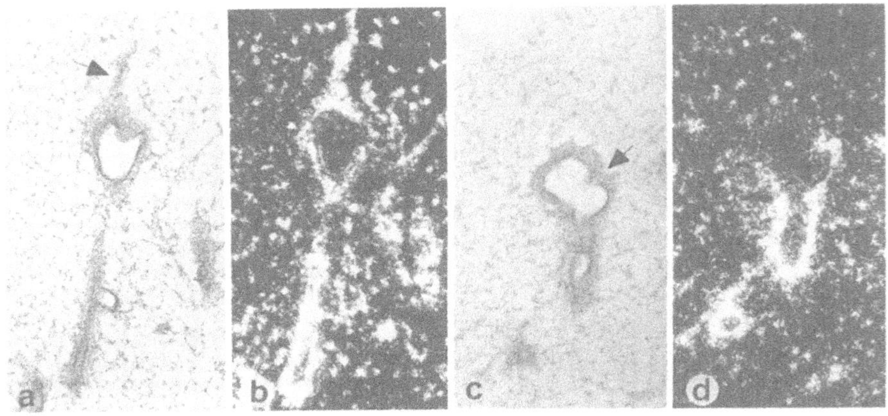

Fig. 8. Localization of type I collagen mRNA by in situ hybridization. *a*, *b*: Light- and dark-field micrographs of an end bud in the vicinity of a TGFβ_1 implant for 24 h. Cellular connective tissue (arrow) and heavy hybridization are apparent over the tip of the end bud. *c*, *d*: Light- and dark-field micrographs of an untreated end bud. There is no labeling over the tip except where a cleft is forming (arrow).

form of a mass at the end bud tip consisting of cells engaged in active transcription of collagen genes, as demonstrated by hybridization in situ with an RNA probe for the rat α1(I) collagen gene (Fig. 8). In control preparations, in which the end buds were regressing because of normal inhibitory mechanisms, collagen mRNA was not detected in unusual amounts at the end bud tip. In both treated and untreated gland, collagen gene transcription was active at the end bud flank and along the subtending ducts, as expected. We also showed that collagen was synthesized in these regions and that another matrix component, sulfated glycosaminoglycans, was undergoing synthesis in the same areas (26).

Perhaps the most interesting aspect of these studies was the observation that stimulation of ECM was epithelium specific. TGFβ_1 itself, in the absence of mammary end bud epithelium, displayed no matrix activity. This was shown by examination of the stroma immediately surrounding the TGFβ_1-containing EVAc implants that in the absence of epithelium, stimulated neither collagen gene expression, collagen production, nor the synthesis of glycosaminoglycans. This indicates that mammary epithelium modifies the contiguous stroma, making it susceptible to TGFβ_1 induction. We postulate that end buds produce an unknown diffusible factor that stimulates matrix synthesis and deposition, which acts in concert with glandular TGFβ_1 or with exogenous TGFβ_1 released from implants. In the latter case, the unusually high levels of TGFβ_1 that might be present around an implant could account for the intense gene expression and ECM synthesis observed at the end bud tip.

In any case, the observation that TGFβ$_1$ stimulates ECM synthesis in an epithelium-specific manner suggests that naturally occurring TGFβ$_1$ also stimulates mammary-associated ECM synthesis. This ECM, which immunostaining shows to be a site of TGFβ$_1$ attachment, may serve to suppress budding and to maintain the normal pattern of branching (see Fig. 7).

CONCLUSION

We have described studies aimed at elucidating the physiological role of TGFβ$_1$ in mammary gland morphogenesis. Using slow-release implants to administer the growth factor directly to local areas of the gland, TGFβ$_1$ was found to be a potent inhibitor of ductal, but not alveolar, growth. This inhibition displayed a number of characteristics, such as rapidity of action, reversibility, and tissue selectivity, that would be expected of a naturally occurring regulatory molecule. TGFβ$_1$ gene transcription was found to be present in the gland, and both mature and latent forms of the growth factor were identified by immunostaining. A mechanism for TGFβ$_1$ inhibition is suggested in which exogenous TGFβ$_1$ stimulates extracellular matrix synthesis and deposition in an epithelium-specific manner. This activity is related to ongoing inductive tissue interactions between mammary epithelium and its surrounding stroma. The fibrous matrix forms a complex with TGFβ$_1$, and perhaps with other developmentally active factors, to suppress further growth of the gland.

REFERENCES

1. Topper YJ, Freeman CS. Multiple hormone interactions in the developmental biology of the mammary gland. Physiol Rev 1980;60:1049-106.
2. Faulkin LJ Jr, DeOme KB. Regulation of growth and spacing of gland elements in the mammary fat pad of the C3H mouse. J Natl Cancer 1960;24:953-69.
3. DeOme KB, Faulkin LJ Jr, Bern HA. Development of mammary tumors from hyperplastic alveolar nodules transplanted into gland-free mammary fat pads of female C3H mice. Cancer Res 1959;19:515-20.
4. Daniel CW. Finite growth span of mouse mammary gland serially propagated in vivo. Experientia 1973;29:1422-4.
5. Daniel CW, Silberstein GB. Postnatal development of the rodent mammary gland. In: Neville MC, Daniel CW, eds. The mammary gland: development, regulation, and function. New York: Plenum Press, 1987.
6. Pitelka DR, Hamamoto ST. Form and function in mammary epithelium: the interpretation of ultrastructure. J Dairy Sci 1977;60:643-54.
7. Williams JM, Daniel CW. Mammary ductal elongation: differentiation of myoepithelium during branching morphogenesis. Dev Biol 1983;97:274-90.
8. Daniel CW, Silberstein GB, Strickland P. Direct action of 17 beta-estradiol on mouse mammary ducts analyzed by sustained release implants and steroid autoradiography. Cancer Res 1987;47:6052-7.
9. Coleman S, Silberstein GB, Daniel CW. Ductal morphogenesis in the mouse mammary gland: evidence supporting a role for epidermal growth factor. Dev Biol 1988; 127:304-15.
10. Daniel CW, Silberstein GB, Van Horn K, Strickland P, Robinson S. TGF-beta-1-

induced inhibition of mouse mammary ductal growth: Developmental specificity and characterization. Dev Biol 1989;135:20-30.
11. Silberstein GB, Daniel CW. Investigation of mouse mammary ductal growth regulation using slow-release plastic implants. J Dairy Sci 1987;70:1981-90.
12. Daniel CW, Silberstein GB. Local effects of growth factors. In: Lippman ME, Dickson RB, eds. Breast cancer: cellular and molecular biology. Boston: Kluwer Academic, 1989.
13. Roberts AB, Sporn MB. The transforming growth factor-betas. In: Sporn MB, Roberts AB, eds. Peptide growth factors and their receptors. Heidelberg: Springer-Verlag, 1989.
14. Silberstein GB, Daniel CW. Elvax 40P implants: sustained, local release of bioactive molecules influencing mammary ductal development. Dev Biol 1982;93:272-8.
15. Silberstein GB, Strickland P, Trumpbour V, Coleman S, Daniel CW. In vivo, cAMP stimulates growth and morphogenesis of mouse mammary ducts. Proc Natl Acad Sci USA 1984;81:4950-4.
16. Vonderhaar BK. Local effects of EGF, TGF, and EGF-like growth factors on lobuloalveolar development of the mouse mammary gland in vivo. J Cell Phys 1987; 132:581-4.
17. Silberstein GB, Daniel CW. Reversible inhibition of mammary gland growth by transforming growth factor-beta. Science 1987;237:291-3.
18. Coleman S, Daniel CW. Inhibition of mouse mammary ductal morphogenesis and down regulation of the EGF receptor by epidermal growth factor. Dev Biol 1990; 137:425-33.
19. Roberts AB, Anzano MA, Wakefield LM, Roche NS, Stern DF,Sporn MB. Type beta transforming growth factor: A bifunctional regulator of cellular growth. Proc Natl Acad Sci USA 1985;82:119-23.
20. Heine UI, Munoz EF, Flanders KC, et al. Role of transforming growth factor beta in the development of the mouse embryo. J Cell Biol 1987;105:2861-76.
21. Akhurst RJ, Fee F, Balmain A. Localized production of TGF-beta mRNA in tumor promoter-stimulated mouse epidermis. Nature 1988;331:363-5.
22. Fava RA, McClure DB. Fibronectin-associated transforming growth factor. J Cell Phys 1987;131:184-9.
23. Rizzino A. Transforming growth factor beta: multiple effects on cell differentiation and extracellular matrices. Dev Biol 1988;130:411-22.
24. Silberstein GB, Daniel CW. Glycosaminoglycans in the basal lamina and the extracellular matrix of serially aged mouse mammary ducts. Mech Age Dev 1984;24: 151-62.
25. Silberstein GB, Daniel CW. Glycosaminoglycans in the basal lamina and extracellular matrix of the developing mouse mammary duct. Dev Biol 1982;90:215-22.
26. Silberstein GB, Daniel CW, Coleman S, Strickland P. Epithelium-dependent induction of mouse mammary gland extracellular matrix by TGF-beta-1. J Cell Biol (in press).

10

Growth Factors as Local Regulators of Normal and Malignant Human Mammary Epithelium

Robert B. Dickson

Vincent T. Lombardi Cancer Center, Georgetown University Hospital, Washington, D.C.

The mechanisms regulating growth of normal and malignant mammary epithelial cells are poorly understood. However, it is known that hormonal influences are critical in the normal development of the mammary gland. The ovaries (under pituitary control) promote glandular growth and differentiation, while the pituitary itself directly controls lactation. Proliferation of normal mammary epithelium seems to require both estrogen and progesterone, but cellular mitoses occur predominantly in the luteal phase of the menstrual cycle, when progesterone is in highest abundance. This is in striking contrast to the endometrium (which is where cellular mitoses occur predominantly in the follicular phase, when estrogen is "unopposed" by progesterone) (1). The pituitary and ovaries are also required for the development of breast cancer in women. The influences of the ovaries in breast cancer seem to be mediated by estrogen and progesterone (again under pituitary control) (2). In rodent models of carcinogen-induced breast cancer, it has been shown that both progesterone and estrogen seem to be required for tumor formation and for early growth of tumors (3–4). Presumably, the mechanism of interaction of estrogen and progesterone in normal and malignant breast is related to the requirement of estrogen to induce expression of progesterone receptor. However, other types of interactions have not been ruled out. Current controversy also surrounds both progestin and estrogen components of the oral contraceptive as risk factors in developing breast cancer (5). Estrogen and progesterone receptors have been localized to a luminal subpopulation of ductal and lobular epithelial cells in women and rodents. Receptors appear to be absent from terminal end bud epithelial cells. Thus, estrogen and progesterone receptors appear to be present in at least partially differentiated epithelium. It is not yet clear whether these steroid receptor positive cells are precursors to breast cancer, though circumstantial evidence suggests the possibility (6–7). In human breast cancer treatment, both antiestrogen and antiprogestin therapy have been successfully used as adjuvant to surgery.

Recent studies have begun to address mechanisms of action of estrogen and progesterone in promotion and growth of malignancy. Many studies are examining oncogenes and growth factors by their expression as well as by gene transfer in cells or the germ line of intact animals (8–11). In hormone-responsive breast cancer cells, growth stimulation by estrogen is accompanied by an increase in growth stimulatory transforming growth factor alpha (TGFα) production (12–14), whereas growth inhibition of hormone responsive breast cancer cell lines by an antiestrogen is paralleled by augmented secretion of growth inhibitory transforming growth factor beta (TGFβ) (15). In hormone-independent breast cancer cell lines, compared to hormone-dependent breast cancer cells, both of these growth factors, as well as many other growth regulatory peptides, are constitutively produced (16–18). These results are consistent with, but do not prove, a role for growth factors in the expression of a more malignant phenotype and escape from normal hormonal control. It is of special note that milk, the natural secretory product of the mammary epithelial cell, is an extraordinarily rich source of growth factors. The three factors upon which this chapter focuses—TGFα, MDGF-I, and TGFβ—are all found in high quantities in milk (19).

The transforming growth factors derive their name from their ability to reversibly induce the transformed phenotype (initially defined as the capacity for anchorage-independent growth) in certain rodent fibroblasts (reviewed in 20). They are polypeptides that were initially found to be synthesized and secreted by a variety of retrovirally, chemically, or oncogene-transformed human and rodent cell lines (11). Two major classes of structurally and functionally distinct transforming growth factors are TGFα and TGFβ. TGFα and TGFα-like peptides consist of multiple species ranging from apparent molecular masses of 6–35 kD (12) and compete with the structural homolog EGF for binding to the same receptor (21). The TGFβ family consists of several related gene products, each forming 25-kD dimeric species (22). There appears to be a complex pattern of interactions of these species with the TGFβ receptor(s) that has been described as three different molecular weight species. TGFβ, and more recently TGFα, have been found in the urine and pleural and peritoneal effusions of cancer patients (23–25). These growth factors have also been observed in some normal tissues (11, 26).

A more recently described growth factor, mammary derived growth factor I (MDGF-I) has been found in human milk and in conditioned medium from human breast cancer cell lines (27–29). This 62-kD growth factor may also play a role in growth regulation of normal and malignant human mammary epithelium. It has been hypothesized that transformation of cells from normal to malignant may directly result from increased production of growth stimulatory factors or decreased production of growth inhibitory substances, or altered responsiveness to either or both transforming groups of growth factors (reviewed in 11). An important counterpoint to understanding pathways of growth control in human neoplastic cells is a knowledge of growth regulation of the normal cells from which the cancer was derived. To date, this area of investigation in epithelial cells has lagged behind in studies of the influence of growth factors in cancer due to the difficulties involved in culture of normal epithelial cells. However, the recent development of specialized serum-free culture conditions has facilitated the study of growth regulation in normal human keratinocytes (30), normal human bronchial epithelial cells (31), and normal human mammary epithelial cells (32–33).

EGF AND TGFα IN DEVELOPMENT AND CARCINOGENESIS OF THE MAMMARY GLAND

The growth factor EGF appears to be an important regulator both of the proliferation and differentiation of the mouse mammary gland in vivo and of mouse mammary explants in vitro (34–35). EGF is also a required supplement for the clonal growth, in vitro, of normal human mammary epithelial cells (37). However, human breast cancer cells do not require exogenous EGF for continuous growth, although many breast cancer cell lines retain receptors and growth stimulatory responses to EGF (37–38). TGFα, a structural and functional homolog of EGF, can produce essentially the same biological effects in mouse mammary explants and cultured human and mouse mammary epithelial cell lines as EGF (34, 39), but its role in normal or malignant mammary development has not been fully defined. Mouse salivary gland-derived EGF appears to be necessary for spontaneous mammary tumor formation in the mouse model (40), as well as for growth of the tumors once they are formed. EGF can also partially replace estrogen to promote limited tumor growth of a human breast cancer cell line (MCF-7) implanted in nude mice (41). A new member of this growth factor family, termed amphiregulin (42), has also been discovered in a breast cancer cell line treated with a tumor promoter, but its role in normal and malignant proliferation remains to be determined.

The growth factor TGFα has been directly implicated as a modulator of cellular transformation in a number of studies. Overexpression of TGFα following transfection of a human TGFα cDNA expression vector into the immortal, but nontumorigenic, mouse mammary epithelial cell line NOG-8 led to its capacity for anchorage-independent growth (43). In addition, in two of three studies using rodent fibroblasts as recipients for human or rat TGFα cDNA, transformation to full tumorigenicity was also achieved (44–45). In contrast, in a third study, TGFα could induce proliferation, but not full malignant progression tumorigenicity (46). EGF can also act as a transformation-inducing agent (an oncogene) when transfected and overexpressed in rodent fibroblasts (47). Some evidence also suggests that oncogene expression in breast cancer is associated with the level of secretion of TGFα. A direct correlation between the degree of TGFα production, *ras* oncogene expression, and the degree of malignant transformation has been demonstrated in a recent study utilizing a glucocorticoid-inducible point-mutated c-Ha-*ras* construct transfected into immortal mouse mammary epithelial cells (48). In studies of human breast cancer biopsies, TGFα mRNA and protein were detected in 70% or more of the specimens (12, 49) and in approximately 30% of benign breast lesions

(50). Immunoreactive TGFα has been found in fibroadenomas and in 25%–50% of primary human mammary carcinomas (13, 51), and an EGF-related protein of 43 kD has been recently isolated from breast cancer patient urine (52).

ras AND *myc* ONCOGENES IN BREAST CANCER

In the classical studies of rodent fibroblasts transformed by Harvey, Kirsten, or Maloney murine sarcoma viruses (20, 53–55), increased production of "sarcoma growth factor" (SGF) was demonstrated. SGF was later characterized as consisting of the two growth factors and TGFβ. Similarly, increased production of TGFα has been reported following transfection of MCF-7 human breast cancer cells with v-Ha-*ras* (the oncogene of Harvey sarcoma virus) or of mouse mammary epithelial cells by a point-mutated, human c-Ha-*ras* gene (39, 48, 56). The actual incidence of the point mutations of the c-Ha-*ras* and c-Ki-*ras* proto-oncogenes in breast cancer appears to be low; they have been observed so far in only two hormone-independent human breast cancer cell lines (57–58). However, the role of unmutated but overexpressed c-Ha-*ras* proto-oncogene in clinical cases of human breast cancer has not yet been fully clarified. One study indicates a positive correlation between expression of p21 *ras* protein and malignant progression of human breast cancer (59). In another study, no such correlation could be observed, although malignant and dysplastic breast lesions did have elevated levels of p21 *ras* protein compared to normal tissues (60). Neither of these studies addressed the state of mutational activation of the *ras* gene, but studies by several other groups indicate that *ras* activation and some additional event(s) are necessary for neoplastic transformation (reviewed in 61).

The proto-oncogene known as c-*myc* is also of interest in breast cancer. This proto-oncogene can confer immortality to fibroblasts (62) and alter fibroblast responsiveness to growth factors (63–64). In human primary breast cancer, c-*myc* amplification has been reported in 15%–30% of the tumors and found to correlate with poor prognosis (65). Expression of the c-*myc* proto-oncogene under control of the mammary-specific WAP-promoter in transgenic mice gave rise to mammary tumors in more than 80% of the animals (66). In vitro studies have also introduced the c-*myc* gene into immortalized human mammary epithelial cells using an amphotropic retroviral vector. In immortalized mammary epithelial cells transfected either with c-*myc* or SV40T (but not v-ras^{H}), it was observed that the cells could be stimulated to grow in soft agar by either bFGF, EGF, or TGFα (67). These data suggest that c-*myc* might function in early breast cancer lesions to allow growth factors to act aberrantly, by driving transformed growth.

EGF/TGFα RECEPTOR

The potential roles of TGFα or EGF in transformation may also involve alterations in the expression and function of their receptor, the EGF receptor. Clinical evidence for the role of increased expression of the EGF receptor, and its structurally related homolog c-*erb*B-2, in more aggressive and hormone-unresponsive breast cancer has accumulated in recent years (68–71). This is also supported by studies of in vitro cultured primary human breast cancer biopsies (72) and established human breast cancer cell lines (38). The EGF receptor expression (but not c-*erb*B-2 expression) appears to be inversely correlated with expression of the estrogen receptor (68). In contrast to breast cellular transfections with mutated c-Ha-*ras* oncogene, which induces TGFα, transfection with the oncogenic counterpart of c-*erb*B-2 (or *neu*) does not induce TGFα (73). In transfection studies on rodent fibroblasts, overexpression of the EGF receptor can predispose cells to expression of the transformed phenotype upon stimulation by EGF (74–76). Likewise, transfection of rodent fibroblasts with c-*erb*B-2, structurally related to the EGF receptor, but lacking EGF binding capacity, results in transformation (77–78). The ligand for the c-*erb*B-2 proto-oncogene has not been definitively identified at present. A new family member, c-*erb*B-3, has also been recently identified in breast cancer (79); however, its implications for breast cancer biology or prognosis are unknown at present.

While expression of positive-acting proto-oncogenes and oncogenes may play a role in breast cancer, it is nearly certain that loss of expression of tumor-suppressing genes plays a great or even greater role. Evidence of this is the frequent, specific observation of chromosomal rearrangements in breast cancer, the familial nature of breast cancer, and recent isolation of two such genes: retinoblastoma and NM23 (80–83).

TGFβ IN VIVO AND IN VITRO

TGFβ appears to be a multifunctional regulator of cell growth (84–85). Depending on the cell type, the presence of serum, or other growth factors, it can either stimulate or inhibit cellular proliferation. Cells of mesenchymal origin, such as fibroblasts, usually respond to TGFβ by growth stimulation. In contrast, most cells of epithelial origin are inhibited by TGFβ. TGFβ also has many other effects on various cell types, such as induction of differentiation of bronchial epithelial cells and chondrocytes (31, 86) and stimulation of synthesis of extracellular basement membrane (10–11, 85). In fibroblasts it has been proposed that TGFβ exerts its growth-promoting effects via induction of c-*sis* (87). However, this is apparently not the mechanism in mammary epithelium, where TGFβ inhibits proliferation and stimulates c-*sis* production (no PDGF receptors are detected) (88). In studies of rodent endothelial cells and in a breast cancer cell line with overexpression of the EGF receptor (89–90), effects of TGFβ on growth were linked to alterations in EGF receptor-binding characteristics. However, this association was also not supported by other studies in normal and malignant breast epithelial lines (91).

TGFβ mRNA has been found in the majority of human tissue types studied to date, unlike the fairly restricted distribution of TGFα. The relative abundance of TGFβ in platelets (92) and bone (86), as well as its stimulatory action on the production of structural and extracellular matrix components, has led to the hypothesis that TGFβ may have a role in the processes of wound repair and tumor cell metastasis (27). Expression of TGFβ mRNA is more abundant in malignant human breast biopsies than in benign lesions (50); it is also more abundant in phenotypically more aggressive, hormone-independent breast cancer cell lines compared to hormone-dependent breast cancer cells (14, 16–17). In the developing normal mouse mammary gland, exogenous TGFβ is a potent, but reversible, in vivo inhibitor of terminal end bud formation (93). In most in vitro systems, no obvious correlation has been observed between expression of the transformed phenotype and resistance to growth inhibition by TGFβ. In one study rodent fibroblasts transformed by Harvey or Maloney sarcoma viruses demonstrated increased production of TGFβ and reduced TGFβ binding (54), but the growth responses to TGFβ were not examined in the transformed cells. Another study using rodent embryo fibroblasts transformed by the activated c-Ha-*ras* gene showed reduced responsiveness to TGFβ (94), but no information on expression of TGFβ or its receptor was provided. Immortalized human bronchial epithelial cells transformed by activated v-Ki-*ras* showed loss of responsiveness to TGFβ compared to their untransformed counterparts (95), but again, expression of TGFβ or its receptor was not described. In two studies of TGFβ responsiveness comparing established cancer cell lines to normal cell lines derived from the same tissue type, an altered responsiveness to TGFβ was seen in the neoplastic cell lines. The loss of responsiveness correlated with absence of receptors for TGFβ in squamous cell carcinoma cell lines compared to normal human prokeratinocytes (96) and in retinoblastoma-derived cell lines compared to human fetal retinal cells (97). Since the cancer cell types in these studies may not have been derived from the same original stem cell population as the normal tissue to which it was compared, it is difficult to draw any conclusions regarding the role of loss of TGFβ receptors in malignant progression.

TGFα AND EGF RECEPTOR EXPRESSION IN HUMAN MAMMARY EPITHELIAL CELLS

The growth responsiveness of normal, immortalized, and oncogene-transformed human mammary epithelial cells to EGF, TGFα, TGFβ, and MDGF-I has been examined. These studies utilized a series of human mammary epithelial cells derived from histopathologically normal breast tissue obtained from reduction mammoplasty on young, nonpregnant, nonlactating women (98). The normal diploid epithelial cell strains 184, 172, and 161 were capable of growing in vitro for 12–20 passages (50–70 cell doublings) on plastic surfaces before undergoing terminal differentiation. They required medium MCDB 170, containing bovine pituitary extract, EGF, insulin, and hydrocortisone, among other supplements, for rapid proliferation and serial passage. The epithelial origin of these cells has been established by the criteria of immunocytochemical markers and electron microscopy (98). After treatment of 184 cells with benzo-*a*-pyrene, immortalized cell lines were established, cloned (32), and used as recipients for various oncogenes carried in retroviral vectors. The immortalized clone 184A1N4 was transfected with the following oncogenes: v-Ha-*ras* (184A1N4-H); v-*mos* (184A1N4-M); SV40T (184A1N4-T); or both v-Ha-*ras* and SV40T (184A1N4-TH); or both v-Ha-*ras* and v-*mos* (184A1N4-MH) (99). A different clone, 184B5, was transfected with v-Ki-*ras* (184B5-Ki).

The carcinogen-immortalized cells could be propagated in a less-complex medium than the parent 184 cells, utilizing IMEM with 0.5% fetal calf serum (FCS), EGF, insulin, and hydrocortisone. Like the 184 cells, they will not grow under anchorage-independent conditions nor form tumors in nude mice. The oncogene-transformed cells were grown in IMEM with 10% FCS. The 184A1N4-T and 184A1N4-M cells cloned poorly in soft agar and are not tumorigenic. The 184A1N4-H and 184A1N4-MH cells were weakly or moderately tumorigenic, respectively, but neither cell line clones well in soft agar. Phenotypically, the 184A1N4-TH and 184B5-Ki cells were fully transformed in that they grew extensively under anchorage-independent conditions and were highly tumorigenic in nude mice. Neither the normal parental 184 cells nor any of the immortalized oncogene-transformed sublines expressed the estrogen receptor when grown under standard culture conditions. Presumably, the cells that propagate under these in vitro conditions are similar in nature to the basal stem cells of the mammary epithelium.

Since the most obvious initial difference in growth requirements among the 184-derived cells was the capability of the transformed cells to grow in a simple, nondefined medium, the cells' responsiveness to EGF or TGFα was concluded (100). In assays for growth response to EGF/TGFα, the degree of growth factor responsiveness diminished with increased expression of the transformed phenotype. The normal parental 184 cells and the immortalized 184A1N4 cells were dependent on exogenous EGF at clonal densities. In high-density culture, however, the 184 cells grew quite well without EGF. It was hypothesized that the density-related EGF-dependence in the 184 cells is related to crossfeeding and autocrine stimulation of growth at higher cell densities. Among the oncogene-carrying cells, those expressing v-Ha-*ras* showed lack of responsiveness to exogenous EGF or TGFα under both anchorage-dependent and -independent conditions. Conversely, the 184A1N4-M and 184A1N4-T were both growth stimulated by EGF/TGFα. While normally not capable of significant anchorage-independent growth, the 184A1N4-T could be induced to clone in soft agar with up to 7% cloning efficiency by supplementation with either EGF or TGFα (100).

To evaluate whether these differences in cellular responsiveness to exogenous EGF/TGFα related to differences in production of endogenous TGFα, TGFα mRNA expression was measured by Northern analysis of total cellular RNA. No differences were observed in abundance of the expected 4.8-kb TGFα species comparing all 184-derived cells. In addition, all of the cells had TGFα mRNA levels comparable to or greater than those of hormone-independent human breast cancer cell lines. Five other normal, nonimmortalized human breast epithelial cell strains also showed the same high TGFα expression (101). Another recent study has detected TGFα in pregnant mammary epithelium and stroma of rodent and women (102). As assayed either by a TGFα-specific radioimmunoassay (103) or by induction of anchorage-independent cloning of normal rat kidney cells in soft agar (16), all oncogene-expressing cells produced comparable levels of bioactive and immunoreactive TGFα. Using an antibody directed against the ligand-binding domain of the EGF receptor, we have subsequently shown that the 184 cells in high-density culture produce sufficient amounts of TGFα to stimulate their own proliferation through an autocrine mechanism, thereby rendering them relatively independent of exogenous EGF or TGFα (101).

An important difference between the normal human mammary epithelial cells used in this study and normal tissue is that these cells are rapidly proliferating in culture. It appears likely that TGFα production is more closely associated with cellular proliferation rather than serving as a marker for transformation. To test this hypothesis, TGFα mRNA expression was studied by Northern analysis and by in situ hybridization techniques in the nonproliferating organoid preparations from which cultures of normal mammary epithelial cell strains are initiated. TGFα mRNA was either not detectable or expressed at very low levels in the organoids (101). TGFα expression also declined in low-density cultures of the 184 cells upon withdrawal of EGF and bovine pituitary extract from the medium. It was concluded that in human mammary epithelial cells, TGFα expression may be directly coupled to the proliferative state. However, it was not directly associated with oncogene-mediated transformation. Instead, as described below, the response of the cells to EGF/TGFα changed both qualitatively and quantitatively with malignant transformation. It should be noted that some, but by no means all, breast cancer cell lines are also inhibited by an antibody directed against the EGF receptor to interrupt an autocrine loop (104). In particular, an autocrine loop was observed when TGFα expression and an EGF receptor gene amplification occurred in the same cell line (104).

Since the observed phenotypic differences among the 184-derived cells could not be explained on the basis of variations in TGFα production, and since variations in growth factor responsiveness and a possible autocrine loop may occur at the receptor level, we examined EGF receptor expression in the human mammary epithelial cells. We found no significant differences in the levels of expression of two EGF receptor-specific mRNA species, 10 and 5.6 kb, comparing the normal and oncogene-transformed cells using Northern analysis of total cellular RNA. Southern analysis of *Hind* III-digested DNA also failed to show any differences; no EGF receptor gene amplifications or rearrangements were observed. High levels of TGFα, EGF receptor, and detectable expression of c-*erb*B-2 have also been reported in proliferating normal human mammary epithelial cells (105). EGF receptor-binding characteristics were subsequently determined for all the human mammary epithelial cell lines. The 184, 184A1N4, and 184A1N4-M cells had 3 to 4 × 10^5 binding sites per cell. In the 184A1N4 and 184A1N4-M cells, both a high- and a low-affinity EGF-binding component were determined, while only the high-affinity component was observed in the 184 cells. The 184A1N4-T cells had markedly elevated levels of both high- and low-affinity binding components, with a total number of binding sites per cell close to 2×10^6. This observation may be significant since these cells could respond with induction of anchorage-independent growth upon EGF or FGF stimulation. We observed a similar effect with c-*myc* transfection of 184A1N4 cells (67).

Studies by other investigators have demonstrated that overexpression of certain oncogenes in some cells is associated with hypersensitivity to EGF. In these other studies this response usually occurred without concomitant changes in either receptor binding or cellular phenotypic characteristics (e.g., c-*myc* overexpression in rat fibroblasts [106]; c-*myc* overexpression in chicken mesenchymal cells [107]; and pp60$^{c\text{-}src}$ overexpression in rat embryo fibroblasts [108]). The high level of EGF receptor sites/cell in the 184A1N4-T cells was not associated with development of a growth inhibitory response to EGF/TGFα. This contrasts with the response of two other carcinoma cell lines to EGF where EGF receptor overexpression has also been observed (the human breast cancer cell line MDA-MB-468 [109]; and the epidermoid cell line A431 [110]). In these cells the EGF response is biphasic: stimulatory at low concentrations and inhibitory at high concentrations.

All three 184-derived cell lines expressing v-Ha-*ras* had a slight reduction in the number of EGF-binding sites per cell, 1.7 to 3×10^5 and lacked the high-affinity EGF-binding component. These findings are in agreement with several other studies showing a reduction in high-affinity EGF receptor binding in the presence of v-Ha-*ras* or activated c-Ha-*ras* oncogenes (39, 48, 111). Our findings differ from the complete absence of EGF binding originally observed in virus-transformed rodent fibroblasts (20). The mechanism and consequences of loss of the high-affinity binding component of the EGF receptor in mammary cells are not yet defined. It should be noted that there are a number of human breast cancer cell lines that exhibit only a fraction of the number of total EGF-binding sites that were observed in the 184-derived cells, yet that demonstrate growth sensitivity to EGF (37).

MDGF-I AND HUMAN MAMMARY EPITHELIAL CELLS

TGFα may not be the only stimulatory growth factor produced by the mammary epithelium. A new growth factor termed mammary derived growth factor I (MDGF-I) has been recently purified to apparent homogeneity from human milk. The factor has an apparent molecular mass of 62 kD and a pl of 4.8 (28). The factor is a pepsin-sensitive and reducing agent-insensitive protein, and the N-terminal sequence of 18 amino acids shows no homology to any known growth-promoting peptides. An apparently identical factor has been isolated from primary breast cancer and human mammary tumor cells, suggesting that MDGF-I might be an autocrine or paracrine growth factor for breast cancer (28–29).

Studies have begun to evaluate the biological effects of, and binding sites for, MDGF-I on human normal and breast cancer cell lines. At a concentration of 10–25 ng/mL, the factor stimulated the growth of estrogen receptor-positive MCF-7 human breast cancer cells by 50%. In this cell line, MDGF-I also stimulated the synthesis of collagen IV, a basement membrane protein, by 40%. It did not have any effect on estrogen receptor-positive ZR75-1 and T47-D breast cancer cell lines or on receptor-negative MDA-MB-231 breast cancer cells. The factor showed a biphasic effect on the estrogen receptor-negative MDA-MB-231 breast cancer cells. The factor showed a biphasic effect on the estrogen receptor-negative MDA-MB-468 cells at concentrations above 5 ng/mL. The growth of normal human mammary epithelial cells (184-strain) was enhanced by 35% by the addition of the factor, whereas, benzo-*a*-pyrene-immortalized nontumorigenic 184A1N4 human mammary epithelial cells were stimulated by about 60%–70%. However, transformation of these cells by SV40T, v-Ha-*ras*, or v-*mos* desensitized them to MDGF-I. Iodinated MDGF-I binds to moderate-affinity sites on the responsive MCF-7, MDA-MB-468, and 184A1N4 cell lines (Kd = 6×10^{-9}M). Crosslinking of [^{125}I]MDGF-I to binding sites with disuccinimidyl suberate (DSS) followed by SDS gel electrophoresis revealed the presence of a major band of molecular weight of approximately 180–200 kD in MCF-7 and MDA-MB-468 breast cancer cell lines. Labeling of this band was inhibited by excess unlabeled MDGF-I, but not by other growth factors. These data suggest that human mammary epithelial cell lines possess receptors of 120–140 kD in size (27) for MDGF-I, but not for other growth factors. Recent studies have demonstrated that upon ligand stimulation, a protein of approximately 185 kD in size becomes rapidly phosphorylated on tyrosine residue(s). The relationship between the binding protein and phosphoprotein remains to be determined, but it seems possible that the MDGF-I receptor has tyrosine kinase activity that is ligand activated.

TGFβ AND HUMAN MAMMARY EPITHELIAL CELLS

The transition from normal to malignant growth patterns may also involve escape from normal growth inhibitory mechanisms. We therefore decided to study the response of normal, immortalized, and oncogene-transformed human mammary epithelial cells to TGFβ. The principal known inhibitory growth factor for epithelial cells, TGFβ, is known to be a hormonally regulated negative growth factor for a hormone-responsive human breast cancer cell line (15). It has been suggested that TGFβ is most likely an autocrine growth inhibitor for hormone-independent human breast cancer cell lines (24, 112). Also, sensitivity to the growth inhibitory effects of TGFβ is attenuated in rat liver epithelial cells by transfection with an activated v-Ha-*ras* oncogene (113). It has been observed that the normal human breast epithelial cell strain 184 was markedly growth inhibited by TGFβ, while the immortalized subclone 184A1N4 was much less sensitive (9). Under anchorage-dependent growth conditions, the 184A1N4-TH were the least sensitive of all the cells, and in soft agar assays, the 184A1N4-TH were not at all inhibited by TGFβ. The proportion of active to latent TGFβ produced by the cells increased slightly with oncogene transformation, but the amount of total TGFβ produced, and the level of TGFβ mRNA, remained unchanged. TGFβ receptor binding parameters also did not change in comparing the 184A1N4 cells and the oncogene transformants. It was concluded that differential responsiveness to TGFβ was to a certain extent correlated with expression of the transformed phenotype. However, the differential responsivity could not be explained solely on the basis of altered production of endogenous TGFβ or differences in TGFβ receptor-binding characteristics. Rather, modulation of TGFβ inhibition of these human breast epithelial cells apparently occurred at a level distal to the TGFβ receptor.

In many other systems, such as in human bronchial epithelial cells, growth inhibition by TGFβ is associated with an induction of differentiation (31). The effect of TGFβ on the expression of epithelial membrane antigen, a derivative of the breast epithelial specific-marker milk-fat globule protein (97), was studied in the 184 cells and sublines. A 10- to 15-fold increase in milk-fat globule protein was induced in the normal cells following TGFβ treatment. Full malignant transformation with SV40T and v-*ras*^{H} oncogenes did not significantly compromise the TGFβ induction of this differentiation antigen (114). Sodium butyrate, another differentiation-inducing agent, was also found to stimulate milk-fat globule protein expression. Thus, the molecular pathway for the growth inhibitory action of TGFβ on the human mammary epithelial cell lines is apparently separable from induction of cellular differentiation, at least as measured by the breast epithelium specific-marker milk-fat globule protein.

It has been postulated that in fibroblasts, the mitogenic action of TGFβ is coupled to, or mediated by, an induction of expression of c-*sis*, the platelet-derived growth factor B-chain (87). We have previously reported the production of both the A and B chains of PDGF by human breast cancer cells (115). In examining three of the normal human mammary epithelial cell strains, we found that they produced PDGF receptor-competing activity in amounts similar to the levels determined in the conditioned media from the human breast cancer cells (88). RNase protection analysis of total cellular RNA revealed strong expression of the PDGF A-chain. Expression of the PDGF B-chain was reversibly induced by TGFβ treatment, while PDGF A-chain mRNA was not affected. At this point the mechanism(s) behind this induction of PDGF B-chain, associated with both growth inhibition by TGFβ in the human mammary epithelial cells and growth stimulation in fibroblasts, remains to be determined. Two other members of the TGFβ family also exist: TGFβ_2 and TGFβ_3 (116). They appear to have similar activities to TGFβ in inhibition of breast epithelial cells (112).

PARACRINE INTERACTIONS IN BREAST CANCER

It is likely that consideration of the autocrine and endocrine interactions of hormones in breast cancer will not be sufficient to allow understanding of growth regulation. Normal mouse mammary epithelial cells in culture require the presence of a stromal component to allow effects of estrogen on proliferation (117). This condition implies a dynamic communication between the two basic cell types in the breast. This has been clearly demonstrated now for preneoplastic epithelial cells in several studies using either fibroblasts or adipocytes (67, 117) to induce their growth. Several growth factors are now likely candidates for these interactions: insulin-like growth factors (118–122), a TGFα-like growth factor (67), and members of the FGF family (118, 123–125). A likely candidate for epithelial modulation of stromal function is TGFβ, which induces stromal proliferation and production of tenascin, a cancer-associated basement membrane component (126).

CONCLUSIONS

From these studies of normal and oncogene-transformed human mammary epithelial cells, it can be concluded that some of the original hypotheses about mechanisms for malignant transformation, based on studies using fibroblasts, may need to be revised with respect to mammary carcinogenesis. TGFα, initially thought to be directly associated with the transformed state, may be more tightly coupled to cellular proliferation, regardless of the state of transformation. This is also supported by the finding of immunoreactive TGFα and TGFα mRNA in benign human breast lesions (13, 50, 127). Likewise, in chemically induced rat mammary adenocarcinomas, the primary lesions were found to express TGFα, whereas the serially transplantable, more progressed carcinomas expressed little or no TGFα (128). This lends additional support to the contention that enhanced TGFα production per se can not solely account for the transformed phenotype in mammary malignancies. However, the TGFα-EGF receptor system may be critical in driving proliferation of early, well-differentiated malignant lesions. Similar conclusions are drawn in a recent study of mouse epidermal or papilloma cells expressing a human TGFα cDNA in skin grafts on mice. TGFα produced by either tumor cells or adjoining normal cells could stimulate tumor growth, but TGFα could not directly influence tumor progression (129).

The notion that neoplastic proliferation involves escape from normal growth regulation, such as inhibition by TGFβ, is, at least partially, supported by the findings that the most fully transformed, tumorigenic members of the 184-derived cells are also least responsive to TGFβ. Clearly, however, the presence of activated oncogenes in these cells causes more profound alterations in cell growth control than can be entirely explained by altered sensitivity to, or production of, TGFβ or TGFα.

Similarly, since oncogenic transformation apparently caused only minor changes in EGF receptor expression and no differences in TGFβ receptor expression in these human mammary epithelial cells, it can be concluded that modulation of cellular responsiveness to these growth factors can occur at a level beyond direct ligand-receptor interactions. Many questions regarding receptor functionality and second-messenger mechanisms have not yet been addressed in our studies of these cells.

Taken together, then, the effects of malignant transformation on cellular responsiveness to two of three major growth regulatory peptides—TGFα, MDGF-I, and TGFβ—can clearly be observed in a model system of normal, immortalized, and oncogene-transformed human mammary epithelial cells. It is apparent that alterations of cellular responsiveness are effects of perturbations of growth control at a level beyond alterations in TGFα or TGFβ production or cell surface receptors. These alterations encompass sensitization to transforming effects of growth factors and complete replacement of the growth factor pathway with an oncogenic mechanism.

REFERENCES

1. Anderson TJ, Battersby S, King RJB, McPherson K, Going JJ. Oral contraceptive use influences resting breast proliferation. Human Pathol 1989;20:1139-44.
2. Welsch CW. Host factors affecting the growth of carcinogen-induced rat mammary carcinomas: A review and tribute to Charles Brenton Huggins. Cancer Res 1985; 45:3415-43.

3. Jabara AG, Toyne PH, Harcourt AG. Effects of time and duration of progesterone administration on mammary tumors induced by DMBA in Sprague Dawley rats. Br J Cancer 1973;27:63-71.
4. Robinson SP, Jordan VC. Reversal of the antitumor effects of tamoxifen by progesterone in the DMBA-induced rat mammary carcinoma model. Cancer Res 1987; 47:5386-90.
5. McCarty KS. Proliferative stimuli in the normal breast: Estrogens or progestins. Human Pathol 1989;20:1137-8.
6. Daniel CW, Silberstein GA, Strickland P. Direct action of 17β estradiol in mouse mammary ducts analyzed by sustained release implants and steroid autoradiography. Cancer Res 1987;47:6052-7.
7. Dulbecco R. Experimental studies in mammary development and cancer: Relevance to human cancer. Adv Oncol 1990;5:3-6.
8. Paul D, Schmidt GH. Immortalization and malignant transformation of differentiated cells by oncogenes in vitro and in transgenic mice. Crit Rev Oncogen 1989;1: 307-21.
9. Heldin CH, Westermark B. Growth factors: Mechanism of action and relations to oncogenes. Cell 1984;37:9-20.
10. Goustin AS, Leof EB, Shipley GD, Moses L. Growth factors and cancer. Cancer Res 1986;46:1015-29.
11. Sporn MB, Roberts AB. Peptide growth factors and inflammation, tissue repair, and cancer. J Clin Inv 1986;78:329-32.
12. Bates SE, Davidson NE, Valverius EM, et al. Expression of transforming growth factor alpha and its mRNA in human breast cancer: Its regulation by estrogen and its possible functional significance. Mol Endocrinol 1988;2:543-55.
13. Perroteau I, Salomon D, DeBortoli M, et al. Immunological detection and quantitation of alpha transforming growth factors in human breast carcinoma cells. Breast Cancer Res Treat 1986;7:201-10.
14. King RJB, Wang DY, Daley RJ, Darbre PD. Approaches to studying the role of growth factors in the progression of breast tumors from the steroid sensitive to insensitive state. J Steroid Biochem 1989;34:133-8.
15. Knabbe C, Wakefield L, Flanders K, et al. Evidence that TGF beta is a hormonally regulated negative growth factor in human breast cancer. Cell 1987;48:417-28.
16. Bates SE, McManaway ME, Lippman ME, Dickson RB. Characterization of estrogen responsive transforming activity in human breast cancer cell lines. Cancer Res 1986;46:1707-13.
17. Artega CL, Tandon AK, Von Hoff DD, Osborne CK. Transforming growth factor β: Potential autocrine growth inhibitor of estrogen receptor-negative human breast cancer cells. Cancer Res 1988;48:3898.
18. Dickson RB, Lippman ME. Control of human breast cancer by estrogen, growth factors, and oncogenes. In: Lippman ME, Dickson RB, eds. Breast cancer: Cellular and molecular biology. Boston: Kluwer Academic, 1988:119-66.
19. Salomon DS, Kidwell WR. Tumor associated growth factors in malignant rodent and human mammary epithelial cells. In: Lippman ME, Dickson RB, eds. Breast cancer: Cellular and molecular biology. Boston: Kluwer Academic, 1988:363-90.
20. Todaro GJ, Marquardt H, Twardzik DR, Reynolds FH, Stephenson JR. Transforming growth factors produced by viral-transformed and human tumor cells. In: Weinstein IB, Vogel HJ, eds. Genes and proteins in oncogenesis. New York: Academic Press, 1983.

21. Massague J. Epidermal growth factor-like transforming growth factor. J Biol Chem 1983;258:13606-13.
22. Cheifetz S, Bassols A, Stanley K, Ohta M, Greenberger J, Massague J. Heterodimeric transforming growth factor β. J Biol Chem 1988;263:10783.
23. Stromberg K, Hudgins R, Orth DN. Urinary TGFs in neoplasia: Immunoreactive TGF-α in the urine of patients with disseminated breast carcinoma. Biochem Biophys Res Comm 1987;144:1059.
24. Artega CL, Hanauske AR, Clark GM, et al. Immunoreactive alpha transforming growth factor (IrαTGF) activity in effusions from cancer patients: A marker of tumor burden and patient prognosis. Cancer Res 1988;48:5023.
25. Sairenji M, Suzuki K, Murakami K, Motohashi H, Okamoto T, Umeda M. Transforming growth factor activity in pleural and peritoneal effusions from cancer and non-cancer patients. Jpn J Cancer Res (Gann) 1987;78:814.
26. Derynck R. Transforming growth factor α. Cell 1988;54:593-5.
27. Bano M, Kidwell WR, Lippman ME, Dickson RB. Characterization of MDGF-1 receptor in human mammary epithelial cell liver. J Biol Chem 1990;265:1874-80.
28. Bano M, Salomon DS, Kidwell WR. Purification of mammary derived growth factor 1 (MDGF 1) from human milk and mammary tumors. J Biol Chem 1985;260:5745-52.
29. Bano M, Lupu R, Kidwell WR, Lippman ME, Dickson RB. Characterization of MDGF1 and its receptor in human breast cancer cells. Proc Am Assoc Cancer Res Washington, D.C., 1990 (in press).
30. Coffey RJ, Derynck R, Wilcox JN, et al. Production and auto-induction of transforming growth factor-α in human keratinocytes. Nature 1987;328:817-20.
31. Masui T, Wakefield LM, Lechner JF, La Veck MA, Sporn MB, Harris CC. Type β transforming growth factor is the primary differentiation-inducing serum factor for normal human bronchial epithelial cells. Proc Natl Acad Sci USA 1986;83:2438-42.
32. Stampfer MR, Bartley JC. Induction of transformation and continuous cell lines from normal human mammary epithelial cells after exposure to benzo-*a*-pyrene. Proc Natl Acad Sci USA 1985;82:2394-8.
33. Hammond SL, Ham RG, Stampfer MR. Serum-free growth of human mammary epithelial cells: Rapid clonal growth in defined medium and extended serial passage with pituitary extract. Proc Natl Acad Sci USA 1984;81:5435-9.
34. Vonderhaar BK. Regulation of development of the normal mammary gland by hormones and growth factors. In: Lippman ME, Dickson RB, eds. Breast cancer: Cellular and molecular biology. Boston: Kluwer Academic, 1988:251-66.
35. Oka T, Tsutsumi O, Kurachi H, Okamoto S. The role of epidermal growth factor in normal and neoplastic growth of mouse mammary epithelial cells. In: Lippman ME, Dickson RB, eds. Breast cancer: Cellular and molecular biology. Boston: Kluwer Academic, 1988:343-62.
36. Stampfer MR. Isolation and growth of human mammary epithelial cells. J Tiss Cult Meth 1985;9:107-15.
37. Osborne CK, Hamilton B, Titus G, Livingston RB. Epidermal growth factor stimulation of human breast cancer cells in culture. Cancer Res;40:2361-6.
38. Davidson NE, Gelmann EP, Lippman ME, Dickson RB. Epidermal growth factor receptor gene expression in estrogen receptor-positive and negative human breast cancer cell lines. Mol Endocrinol 1987;1:216-23.
39. Salomon DS, Perroteau I, Kidwell WR, Tam J, Derynck R. Loss of growth respon-

siveness to epidermal growth factor and enhanced production of alpha-transforming growth factors in *ras*-transformed mouse mammary epithelial cells. J Cell Physiol 1987;130:397-409.

40. Kurachi H, Okamoto S, Oka T. Evidence for the involvement of the submandibular gland epidermal growth factor in mouse mammary tumorigenesis. Proc Natl Acad Sci USA 1985;81:5940-3.
41. Dickson RB, McManaway ME, Lippman ME. Estrogen-induced factors of breast cancer cells partially replace estrogen to promote tumor growth. Science 1986; 232:1540-3.
42. Shoyab M, Plowman GD, McDonald VL, Bradley JG, Todaro GJ. Structure and function of human amphiregulin: A member of the epidermal growth factor family. Science 1989;243:1074-6.
43. Shankar V, Ciardiello F, Kim N, et al. Transformation of normal mouse mammary epithelial cells following transfection with a human transforming growth factor alpha cDNA. Mol Carcinog 1989.
44. Rosenthal A, Lindquist PB, Bringman TS, Goeddel DV, Derynck R. Expression in rat fibroblasts of a human transforming growth factor-α cDNA results in transformation. Cell 1986;46:301-9.
45. Watanabe S, Lazar E, Sporn MB. Transformation of normal rat kidney (NRK) cells by an infectious retrovirus carrying a synthetic rat type α transforming growth factor gene. Proc Natl Acad Sci USA 1987;84:1258-62.
46. Finzi E, Fleming T, Segatto O, et al. The human transforming growth factor type α coding sequence is not a direct-acting oncogene when overexpressed in NIH 3T3 cells. Proc Natl Acad Sci USA 1987;84:3733-7.
47. Stern DF, Hare DL, Cecchini MA, Weinberg RA. Construction of a novel oncogene based on synthetic sequences encoding epidermal growth factor. Science 1987; 235:321-4.
48. Ciardiello F, Kim N, Hynes N, et al. Induction of transforming growth factor α expression in mouse mammary epithelial cells after transformation with a point-mutated c-Ha-*ras* protooncogene. Mol Endocrinol 1988;2:1202-16.
49. Gregory H, Thomas CE, Willshire IR, et al. Epidermal and transforming growth factor α in patients with breast tumors. Br J Cancer 1989:605-9.
50. Travers MR, Barrett-Lee PJ, Berger U, et al. Growth factor expression in normal, benign, and malignant breast tissue. Br Med J 1988;296:1621-30.
51. Macias A, Perez R, Hägerström T, Skoog L. Identification of transforming growth factor alpha in human primary breast carcinomas. Anticancer Res 1987;7:1271-80.
52. Eckert K, Granetzny A, Fischer J, Nexo E, Grosse R. An Mr 43,000 epidermal growth-factor related protein purified from the urine of breast cancer patients. Cancer Res 1990;50:642-7.
53. Sporn MB, Todaro GJ. Autocrine secretion and malignant transformation of cells. N Eng J Med 1980;303:878-80.
54. Anzano MA, Roberts AB, De Larco JE, et al. Increased secretion of type β transforming growth factor accompanies viral transformation of cells. Mol Cell Biol 1985; 5:242-50.
55. Anzano MA, Roberts AB, Smith JM, Sporn MB, DeLarco JE. Sarcoma growth factor from conditioned medium of virally transformed cells is composed of both type α and type β transforming growth factors. Proc Natl Acad Sci USA 1983;80:6264-8.

56. Dickson RB, Kasid A, Huff KK, et al. Activation of growth factor secretion in tumorigenic states of breast cancer induced by 17-β-estradiol or v-*ras*^{H} oncogene. Proc Natl Acad Sci USA 1987;84:837-41.
57. Kraus MH, Yuspa Y, Aaronson SA. A position 12-activated H-*ras* oncogene in all Hs578T mammary carcinosarcoma cells but not normal mammary cells of the same patient. Proc Natl Acad Sci USA 1984;81:5384-8.
58. Kozma SC, Bogaard ME, Buser K, et al. The human c-Kirsten *ras* gene is activated by a novel mutation in codon 13 in the breast carcinoma cell line MDA-MB 231. Nucleic Acids Res 1988;15:5963-71.
59. Clair T, Miller WR, Cho-Chung YS. Prognostic significance of the expression of a *ras* protein with a molecular weight of 21,000 by human breast cancer. Cancer Res 1987;47:5290.
60. Horan-Hand P, Vilase V, Thor A, Ohuchi N, Schlom J. Quantitation of Harvey *ras* p21 enhanced expression in human breast and colon carcinomas. J Natl Cancer Inst 1987;79:59-65.
61. Medina D. The preneoplastic state in mouse mammary tumorigenesis, Carcinogenesis 1988;9:1113-20.
62. Kelekar A, Cole MD. Immortalization by c-*myc*, H-*ras*, and Ela oncogenes induces differential cellular gene expression and growth factor responses. Mol Cell Biol 1987;7:3899-907.
63. Leof EB, Proper JA, Moses HL. Modulation of transforming growth factor type β action by activated *ras* and c-*myc*. Mol Cell Biol 1987;7:2649-52.
64. Stern DF, Roberts AB, Roche NS, Sporn MB, Weinberg RA. Differential responsiveness of *myc*- and *ras*-transfected cells to growth factors: Selective stimulation of *myc*-transfected cells by epidermal growth factor. Mol Cell Biol 1986;6:870-7.
65. Escot C, Theillet C, Lidereau R, et al. Genetic alteration of the c-*myc* proto-oncogene in human primary breast carcinomas. Proc Natl Acad Sci USA 1986;83:4834-8.
66. Schoenberger CA, Andres AC, Groner B, van der Valk M, LeMeur M, Gerlinger P. Targeted c-*myc* gene expression in mammary glands of transgenic mice induces mammary tumors with constitutive mild protein gene transcription. EMBO J 1988;7:169-75.
67. Valverius EM, Ciardiello F, Kim N, et al. Basic fibroblast growth factor (bFGF) or cocultivation with mammary fibroblasts can induce a transformed phenotype in vitro in immortalized SV40-T expressing human mammary epithelial cells. Proc 5th annu meet Oncogenes, Frederick, MD, 1989:230.
68. Sainsbury JR, Farndon JR, Needham GK, Malcolm AJ, Harris AL. Epidermal-growth-factor receptor status as predictor of early recurrence of and death from breast cancer. Lancet 1987;i:1398-402.
69. Perez R, Pascual M, Macias A, Lage A. Epidermal growth factor receptors in human breast cancer. Breast Cancer Res Treat 1984;4:189-93.
70. Slamon DJ, Godulphin W, Jones LA, et al. Studies of the HER-2/neu protooncogene in human breast and ovarian cancer. Science 1989;244:621-4.
71. Paik S, Hazan R, Fisher ER, et al. Pathologic findings from the National Surgical Adjuvant breast and bowel project: Prognostic significance of erbB$_2$ protein overexpression in primary breast cancer. J Clin Oncol 1990;8:103-12.
72. Spitzer E, Grosse R, Kunde D, Schmidt HE. Growth of mammary epithelial cells in breast-cancer biopsies correlates with EGF binding. Int J Cancer 1987;39:279-82.

73. Ciardiello F, Hynes N, Kim N, Valverius EM, Lippman ME, Salomon DS. Transformation of mouse mammary epithelial cells with the Ha-*ras* but not the neu oncogene results in a gene dosage-dependant increase in transforming growth factor α production. FEBS Lett 1979;250:474-8.
74. Velu TJ, Beguinot L, Vass WC, et al. Epidermal growth factor-dependent transformation by a human EGF receptor proto-oncogene. Science 1987;238:1408-50.
75. Di Fiore PP, Pierce JH, Fleming TP, et al. Overexpression of the human EGF receptor confers an EGF-dependent transformed phenotype to NIH 3T3 cells. Cell 1987;51:1063-70.
76. Riedel H, Massoglia S, Schlessinger J, Ullrich A. Ligand activation of overexpressed epidermal growth factor receptors transforms NIH 3T3 mouse fibroblasts. Proc Natl Acad Sci USA 1988;85:1477-82.
77. Hudziak RM, Schlessinger J, Ullrich A. Increased expression of the putative growth factor receptor $p185^{HER2}$ causes transformation and tumorigenesis of NIH 3T3 cells. Proc Natl Acad Sci USA 1987;84:7159-62.
78. Di Fiore PP, Pierce JH, Kraus MH, Segatto O, King CR, Aaronson SA. *erb*B-2 is a potent oncogene when overexpressed in NIH/3T3 cells. Science 1987b;237:178-82.
79. Kraus MH, Issing W, Miki T, Popescu NC, Aaronson SA. Isolation and characterization of ERBB3, a third member of the ERB/epidermal growth factor receptor family: Evidence for overexpression in a subset of human mammary tumors. Proc Natl Acad Sci USA 1989;86:9193-7.
80. Ali I, Lidereau R, Thillet E, Callahan R. Reduction to homozygosity of genes on chromosome 11 in human breast neoplasia. Science 1987;238:185-7.
81. Lundberg C, Skoog L, Cavence W, Nordenskjold M. Loss of heterozygosity in human ductal breast tumors indicates a recessive mutation on chromosome 13. Proc Natl Acad Sci USA;84:2372-6.
82. Steeg PS, Bevilacqua G, Kopper L, et al. Evidence for a novel gene associated with low tumor metastatic potential. J Natl Cancer Inst 1988;80:200-4.
83. Lee EY, To H, Shew JY, Sculley P, Lee WH. Inactivation of the retinoblastoma susceptibility in human breast cancers. Science 1988;241:218-21.
84. Moses HL, Tucker RF, Leof EB, Coffey RJ, Halper J, Shipley GD. Type-β transforming growth factor is a growth stimulator and a growth inhibitor. In: Feramisco J, Ozanne B, Stiles C, eds. Cancer cells; vol 3. Cold Spring Harbor, NY: Cold Spring Harbor Lab., 1985.
85. Roberts AB, Anzano MA, Wakefield LM, Roche NS, Stern DF, Sporn MB. Type β transforming growth factor: a bifunctional regulator of cellular growth. Proc Natl Acad Sci USA 1985;82:119-24.
86. Seyedin SM, Thompson AY, Bentz H, et al. Cartilage-inducing factor A. J Biol Chem 1986;261:5693-9.
87. Leof EB, Proper JA, Shipley GD, Di Corleto PE, Moses HL. Induction of c-*sis* mRNA and activity similar to platelet-derived growth factor by transforming growth factor-beta: A proposed model for indirect mitogenesis involving autocrine activity. Proc Natl Acad Sci USA 1986;83:2453-7.
88. Bronzert DA, Bates SE, Sheridan JA, et al. TGF beta induces PDGF mRNA and PDGF secretion while inhibiting growth in normal human mammary epithelial cells. Mol Endocrinol 1990 (in press).
89. Takehara K, LeRoy EC, Grotendorst GR. TGF-β inhibition of endothelial cell prolif-

eration: Alteration of EGF binding and EGF-induced growth-regulatory (competence) gene expression. Cell 1987;49:415-22.
90. Fernandez-Pol JA, Klos DJ, Hamilton PD, Talkad VD. Modulation of epidermal growth factor receptor gene expression by transforming growth factor-β in a human breast carcinoma cell line. Cancer Res 1987;47:4260-5.
91. Valverius EM, Walker-Jones D, Bates SE, et al. Production of and responsiveness to transforming growth factor-β in normal and oncogene-transformed human mammary epithelial cells. Cancer Res 1989;49:6269-74.
92. Assoian RK, Komoriya A, Meyers CA, Smith DM, Sporn MB. Transforming growth factor β in human platelets: Identification of a major storage site, purification and characterization. J Biol Chem 1983;259:9756.
93. Silberstein GB, Daniel CW. Reversible inhibition of mammary gland growth by transforming growth factor-β. Science 1987;237:2391-5.
94. Leof EB, Proper JA, Moses HL. Modulation of transforming growth factor type β action by activated *ras* and c-*myc*. Mol Cell Biol 1987;7:2649-52.
95. Reddel RR, Ke Y, Kaighn ME, et al. Human bronchial epithelial cells neoplastically transformed by v-Ki-*ras*: Altered response to inducers of terminal squamous differentiation. Oncogene Res 1988.
96. Shipley GD, Pittelkow MR, Wille JJ Jr, Scott RE, Moses HL. Reversible inhibition of normal human prokeratinocyte proliferation by type β transforming growth factor-growth inhibitor in serum-free medium. Cancer Res 1986;46:2068-73.
97. Kimchi A, Wang XF, Weinberg RA, Cheifetz S, Massague J. Absence of TGF-β receptors and growth inhibitory responses in retinoblastoma cells. Science 1988;240:196-9.
98. Stampfer MR, Bartley JC. Human mammary epithelial cells in culture: Differentiation and transformation. In: Lippman ME, Dickson RB, eds. Breast cancer: Cellular and molecular biology. Boston: Kluwer Academic, 1988:1-24.
99. Clark R, Stampfer MR, Milley R, et al. Transformation of human mammary epithelial cells by oncogenic retroviruses. Cancer Res 1988;48:4689-94.
100. Valverius EM, Bates SE, Stampfer MR, et al. Transforming growth factor alpha production and EGF receptor expression in normal and oncogene transformed human mammary epithelial cells. Mol Endocrinal 1989;3:203-14.
101. Bates SE, Valverius EM, Ennis BW, et al. Expression of the transforming growth factor α/epidermal growth factor receptor pathway in normal human breast epithelial cells. Endocrinology 1990;126:596-607.
102. Liscia DS, Merlo G, Ciardiello F, et al. Transforming growth factor-α messenger RNA localization in the developing adult rat and human mammary gland by in situ hybridization. Dev Biol 1990 (in press).
103. Linsley PS, Hargreaves WR, Twardzik DR, Todaro GJ. Detection of larger polypeptides structurally and functionally related to type I transforming growth factor. Proc Natl Acad Sci USA 1985;82:356-60.
104. Ennis BW, Valverius E, Lippman ME, et al. Anti-EGF receptor antibodies inhibit the autocrine stimulated growth of MDA-MB-468 human breast cancer cells. Mol Endocrinol 1989;3:1830-8.
105. Zajchowski D, Band V, Pauzie N, Tager A, Stampfer M, Sager R. Expression of growth factors and oncogenes in normal and tumor-derived human mammary epithelial cells. Cancer Res 1988;48:7041-7.
106. Stern DF, Roberts AB, Roche NS, Sporn MB, Weinberg RA. Differential responsive-

ness of *myc*- and *ras*-transfected cells to growth factors: Selective stimulation of *myc*-transfected cells by epidermal growth factor. Mol Cell Biol 1986;6:870-94.

107. Balk SD, Riley TM, Gunther HS, Morisi A. Heparin-treated, v-*myc*-transformed chicken heart mesenchymal cells assume a normal morphology but are hypersensitive to epidermal growth factor (EGF) and brain fibroblast growth factor (bFGF); cells transformed by the v-Ha-*ras* oncogene are refractory to EGF and bFGF but are hypersensitive to insulin-like growth factors. Proc Natl Acad Sci USA 1985; 82:5781-5.
108. Luttrell DK, Luttrell LM, Parsons SJ. Augmented mitogenic responsiveness to epidermal growth factor in murine fibroblasts that overexpress $pp60^{c\text{-}src}$. Mol Cell Biol 1988;8:497-501.
109. Filmus J, Pollak MN, Cailleau R, Buick RN. MDA-468, a human breast cancer cell line with a high number of epidermal growth factor (EGF) receptors, has an amplified EGF receptor gene and is growth inhibited by EGF. Biochem Biophys Res Commun 1985;128:898-905.
110. Santon JB, Cronin MT, MacLeod CL, Mendelsohn J, Masui H, Gill GN. Effects of epidermal growth factor receptor concentration on tumorigenicity of A431 cells in nude mice. Cancer Res 1986;46:4701-5.
111. Kamata T, Feramisco JR. Is the *ras* oncogene protein a component of the epidermal growth factor receptor system? In: Levine AJ, Vande Woude GF, Topp WC, Watson JD, eds. Cancer Cells 1/the transformed phenotype. Cold Spring Harbor, NY: Cold spring Harbor Lab., 1984.
112. Zugmaier G, Ennis BW, Deschauer B, et al. Transforming growth factors type β1 and β2 are equipotent growth inhibitors of human breast cancer cell lines. J Cell Physiol 1989;141:353-61.
113. Houck, KA, Strom SC, Michalopoulos G. Resistance to the growth inhibitory effect of transforming growth factor beta is induced by transfection of an activated H-*ras* oncogene into rat liver epithelial cells [Abstract 254]. Proc Am Assn Cancer Res 1987;28:64.
114. Walker-Jones D, Valverius EM, Stampfer MS, Lippman ME, Dickson RB. Transforming growth factor beta stimulates expression of epithelial membrane antigen in normal and oncogene-transformed human mammary epithelial cells. Cancer Res 1989;49:6407-11.
115. Bronzert DA, Pantazis P, Antoniades HN, et al. Synthesis and secretion of PDGF-like growth factor by human breast cancer cell lines. Proc Natl Acad Sci USA 1987;84:5763-7.
116. Graycar JL, Miller DA, Arrick BA, Lyons RM, Moses HL, Derynck R. Human transforming growth factor β3: Recombinant expression, purification, and biological activities in comparison with transforming growth factors -β1 and -β2. Mol Endocrinol 1989;3:1977-86.
117. Haslam SZ. Mammary fibroblast influence on normal mouse mammary epithelial responses to estrogen in vitro. Cancer Res 1986;46:310-6.
118. Cullen KJ, Hill S, Paik S, Smith HS, Lippman ME, Rosen N. Growth factor mRNA expression by human breast fibroblasts from benign and malignant lesions. Proc Am Assn Cancer Res, Washington, D.C., 1990.
119. Foekens JA, Portengen H, van Putten WLJ, et al. Prognostic value of receptors for insulin-like growth factor 1, somatostatin, and epidermal growth factor in breast cancer. Cancer Res 1989;49:7002-9.

120. Peyrat JP, Bonmeterre J, Beuscart R, Djiane J, Demaille A. Insulin-like growth factor 1 receptors in human breast cancer and their relation to estradiol and progesterone receptors. Cancer Res 1988;48:6429-33.
121. Yee D, Paik S, Rosen N, Lippman ME, Cullen KJ. Growth regulation of human breast cancer by insulin-like growth factors. In: Lippman ME, Dickson RB, eds. Breast cancer: Cellular and molecular biology; vol 2. Boston: Kluwer Academic, 1990 (in press).
122. Hilf R. The actions of insulin as a hormonal factor in breast cancer. Banbury report 8: Hormones and breast cancer. Cold Spring Harbor Lab., 1981:317-37.
123. Takahashi K, Suzuki K, Kawahara S, Ono T. Growth stimulation of human breast epithelial cells by basic fibroblast growth factor in serum free medium. Int J Cancer 1989;43:870-4.
124. Levay-Young BK, Imagawa W, Wallace DR, Nandi S. Basic fibroblast growth factor stimulates the growth and inhibits casein accumulation in mouse mammary epithelial cells in vitro. Mol Cell Endocrinol 1989;62:327-36.
125. Kern GF, Wellstein A, Flamm S, et al. Secretion of heparin binding growth factors by breast cancer cells and their role in promoting cancer cell growth. Proc 5th Nagoya Int Symposium on Cancer Treatment. Excerpta Medica International Congress Series, Amsterdam, 1990 (in press).
126. Chiquet-Ehrismann R, Kalla P, Pearson CA. Participation of tenascin and transforming growth factor β in reciprocal epithelial-mesenchymal interactions of MCF-7 cells and fibroblasts. Cancer Res 1989;49:4322-5.
127. Ciardiello F, Kim N, Liscia DS, et al. mRNA expression of transforming growth factor alpha in human breast carcinomas and its activity in effusions of breast cancer patients. J Natl Cancer Inst 1989;81:1165-71.
128. Liu SC, Sanfilippo B, Perroteau I, Derynck R, Salomon DS, Kidwell WR. Expression of transforming growth factor α (TGFα) in differentiated rat mammary tumors: Estrogen induction of TGFα production. Mol Endocrinol 1987;1:683-92.
129. Finzi E, Kilkenny A, Strickland JE, et al. TGFα stimulates growth of skin papillomas by autocrine and paracrine mechanisms but does not cause neoplastic progression. Mol Carcinog 1988;1:7-15.

PART

IV

REGULATORY PEPTIDES IN REPRODUCTIVE TRACT DEVELOPMENT AND FUNCTION

11

Mullerian Inhibiting Substance Activity in the Development of the Reproductive Tract and Lung in the Fetus and Control of Oocyte Meiosis in the Adult

David T. MacLaughlin, Tatsuo Kuroda, Elizabeth A. Catlin, and Patricia K. Donahoe

Pediatric Surgical Research Laboratory, Massachusetts General Hospital and Harvard Medical School, Boston

The development of the normal reproductive tract in male embryos is the result of a series of carefully orchestrated anabolic and catabolic events that for the purposes of this discussion, center about the differentiation of the primitive gonad to a testis. This critical process, presumably directed by the testis-determining factor gene(s) (1–2) results, ultimately, in the ability of the gonad to produce two major testicular hormones: testosterone and mullerian inhibiting substance (MIS). It is the activity of these two factors, acting in concert, to direct the maturation of the male reproductive tract from the wolffian duct system and to cause the regression of the female reproductive tract anlage, the mullerian duct. A variety of studies have shown that once the genetic sex of the gonad is established as testicular, MIS, synthesized and secreted by the Sertoli cells, is responsible for mullerian duct regression (3–4). Testosterone can facilitate this activity of MIS, but cannot, on its own, direct this important event in the development of the male phenotype. Although the fetal ovary does not produce MIS, it has recently been shown that the granulosa cells of the young and adult ovary synthesize and secrete bioactive MIS (5–8). The role of MIS in the female is, obviously, quite different from the major role played by MIS in the male, namely, the regression of the mullerian duct. Our evidence points to MIS as an inhibitor of oocyte meiosis, thus contributing to the proper development of mature germ cells (9).

This chapter focuses on the action of MIS in the fetal male and presents important newly appreciated functions of this growth regulatory factor on extramullerian tissues to the extent that they impact on normal growth and development and reproductive function. Among the extra-mullerian target tissues germane to these considerations are the ovarian granulosa cells and oocytes (8–9) and the fetal lung (10).

Central to our understanding of MIS action is the realization that it appears to function as an inhibitor of growth factor-mediated stimulation of specific target tissues. That is, all the effects of MIS on the fetal urogenital ridge (mullerian duct regression), on oocytes (inhibition of meiosis), and fetal lung (inhibition of surfactant accumulation) are tied to the action of epidermal growth factor (EGF).

Mullerian duct regression caused by MIS in vitro is blocked by EGF (11), as is the MIS inhibition of germinal vesicle breakdown of mature oocytes (12). MIS, on the other hand, inhibits EGF-dependent growth of a human vulvar tumor cell line (A431) in soft agar (13). The action of MIS in these cells, and in the fetal lung as well (unpublished results), inhibits EGF-dependent autophosphorylation of tyrosine residues in the EGF receptor (13–14). The role of MIS, therefore, may be to down-regulate growth factor action by blocking receptor autophosphorylation in a very specific manner by binding to its own receptor since MIS does not appear to alter EGF receptor complex formation. Normal male phenotype and normal oocyte development in females would result from a proper ratio of the bioactivity of gonadal steroids, MIS, and EGF. Conversely, deviations from this delicate balance would result in congenital anomalies in males and impaired fertility in females.

ONTOGENY OF MIS PRODUCTION IN MALES AND FEMALES

The existence of MIS was first proposed for fetal males long before any appreciation of the role of this substance in females. The elegant studies of Jost and coworkers nearly a half a century ago revealed that a testicular product other than testosterone was required to rid the developing urogenital ridge of the precursor to the female reproductive tract components, the upper third of the vagina, the cervix, uterus, fallopian tubes, and the lining of the ovary (15–18). This factor, termed mullerian inhibiting substance (MIS) by Jost or anti-mullerian hormone (AMH) by Professor Josso (19), became the subject of intense investigation, particularly in the early 1970s. The first critical turning point in MIS research came with the development of a reliable and highly specific bioassay for MIS employing the 14.5-day fetal rat urogenital ridge as a test tissue and bovine neonatal testis as a source for MIS (20–21). This bioassay allowed for monitoring the purification of MIS (see below) and provided an important tool with which to assess the ontogeny of MIS production by fetal and adult gonads. Early studies performed before an adequate immunoassay existed for MIS used either extracts of gonads or tissue fragments placed in the bioassay to detect the presence of MIS activity. As can be seen in Figure 1, MIS production in the rat testis as detected by immunohistochemical methods is confined to the Sertoli cell (22).

Bioassayable MIS in the rat appears by day 15 in fetal life, and its production extends at maximum levels for several days after birth, thereafter falling to basal levels (22) (Fig. 2).

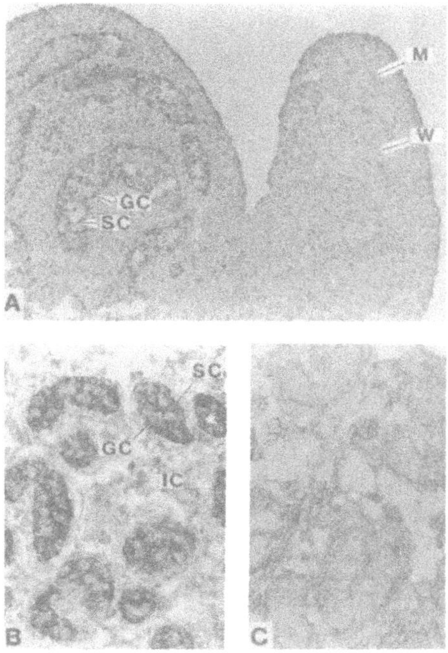

Fig. 1. Avidin-biotin complex method of detecting MIS in the testes using rabbit polyclonal antihuman recombinant MIS antibody (×200). Sertoli cells (SC) stained intensely, while germ cells (GC) and interstitial cells (IC) failed to stain in the testes of both 15-day fetal rat (*A*) and 1-day postnatal rat (*B*); control rabbit IgG did not stain the same cells (*C*). Note the regression of the mullerian duct (M) compared to the wolffian duct (W), indicating functional biological activity of MIS from the adjacent 15-day fetal testis. (From Ueno S, Takahashi M, Manganaro TF, Ragin RC, Donahoe PK, Cellular localization of mullerian inhibiting substance in the developing rat ovary, Endocrinology 1989;124(2):1000–6, with permission.)

Females on the other hand, do not produce bioactive MIS until well after birth (22) (Fig. 3). As mentioned above, the granulosa cells, particularly those surrounding the oocytes, are the source of the hormone (22) (Fig. 4).

More recently, a sensitive immunoassay was developed in our laboratory to measure human MIS in biological fluids (23). Our results on the ontogeny of serum MIS in this species parallel those observed by bioassay in the rat (Fig. 5). That is, males and females demonstrate a sexually dimorphic mode of production early in life, but reach near-equality as adults. We have not yet been able to detect variations in human serum MIS during the normal menstrual cycle, as the ELISA assay now performed lacks the necessary sensitivity. Immunohistochemical studies in

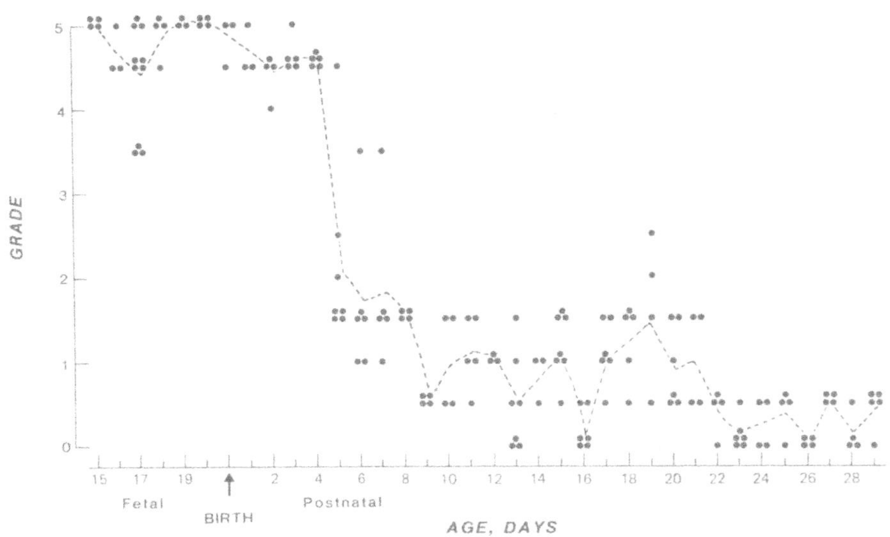

Fig. 2. MIS bioactivity in developing rat testes. The highest MIS activity (grade 5) was detected in fetal testes (15–20 days) and extended for 4 days after birth. Activity declined to intermediate levels over the next 2 weeks and dropped to barely detectable levels (grades 0 to 1) in fragments from testes of animals 22 days of age and older. Each point represents the MIS activity from a single 1-mm fragment from a separate testis. (From Ueno S, Takahashi M, Manganaro TF, Ragin RC, Donahoe PK, Cellular localization of mullerian inhibiting substance in the developing rat ovary, Endocrinology 1989;124(2):1000–6, with permission.)

the rat, however, predict that levels would reach a maximum just before ovulation and decline thereafter, only to rise again with the next cycle (24).

It follows logically that in certain clinical situations, such as tumors of the Sertoli or granulosa cell type, elevated quantities of MIS will be produced and that serum measurements in these patients may be useful. This is, in fact, the case since abnormally high levels of MIS have been detected in patients with granulosa cell tumors and in one patient with a sex cord tumor (23). In the latter case, MIS levels dropped upon cytoreductive surgery, and subsequent rising levels predicted recurrence of the tumor. Whether MIS measurements in children with congenital anomalies of the reproductive tract or respiratory distress (see below) will be of clinical value remains to be established.

MIS PROTEIN AND GENE STRUCTURE

Naturally occurring MIS was purified from neonatal bovine testes using a variety of standard biochemical techniques, including serial column chromatography (25) or

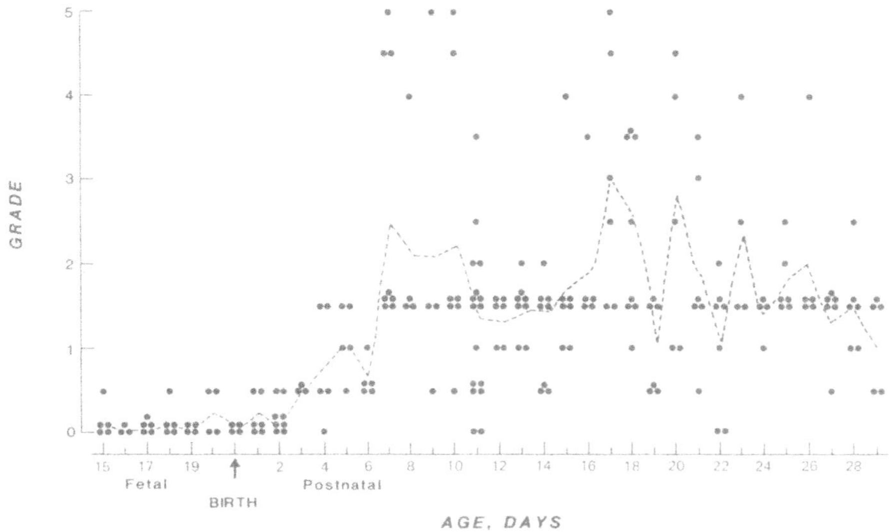

Fig. 3. MIS bioactivity in developing rat ovaries. Ovarian fragments from fetal (15–20 days) and neonatal (0–3 days) rats showed virtually no MIS activity. Little (grades 1–2), but distinctive, MIS activity was detected in the fragments from the 4-day-old rat ovary. Ovarian fragments from rats 7 days of age and older could cause complete mullerian duct regression (grade 5), but extreme individual variations were noted. Each point represents the MIS activity of a single 1-mm fragment from a separate ovary. (From Ueno S, Takahashi M, Manganaro TF, Ragin RC, Donahoe PK, Cellular localization of mullerian inhibiting substance in the developing rat ovary, Endocrinology 1989;124(2):1000–6, with permission.)

immunoaffinity chromatography (26). The resulting preparation, monitored using the in vitro bioassay (21), revealed MIS to be a large glycoprotein homodimer of ~140 kD. The 70-kD monomers are separated from one another by disulfide bond reduction. Early analyses estimated the carbohydrate content of bovine MIS at 12%–18% of the total molecular weight (27). Purified bovine MIS was partially sequenced, and the information used to construct cDNA probes to identify mRNA coding for bovine MIS and, ultimately, to clone the bovine and human MIS gene (28). The human genomic clone contains 5 exons and 4 introns. Sequence analyses of both DNA and the deduced primary structure of MIS revealed significant homology with several other proteins that are all included in a large supergene family, including the transforming growth factor βs (TGFβs), inhibins/activins, and the bone morphogenesis factors in mammals, as well as the Vg_1 protein of *Xenopus laevis* and the decapentaplegia complex of *Drosophila* (28). The area of greatest homology exists between MIS and the TGFβs in the carboxy terminal end of the

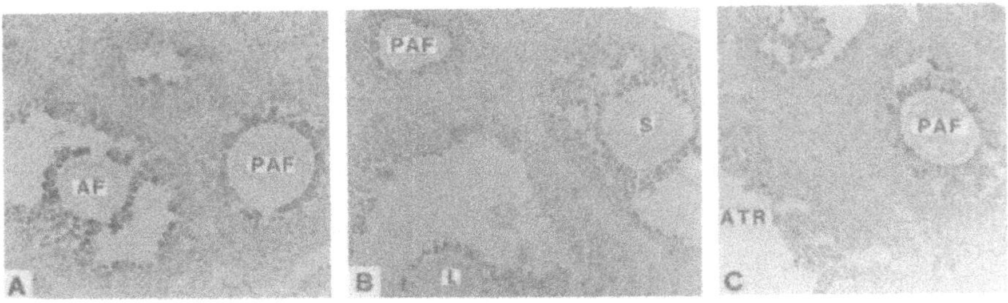

Fig. 4. Immunohistochemical staining (MIS) of the 27-day-old (*A,C*) and 21-day-old (*B*) rat ovary (×200). *A*: The innermost granulosa cell layer of preantral follicles (PAF) stain more intensely than the more peripheral granulosa cells. In the antral follicles (AF), the cumulus cells as well as the inner granulosa cells stain intensely. *B*: Inner granulosa and cumulus cells in larger antral follicles (L) stain less intensely than cells in smaller (S) or preantral follicles (PAF). *C*: In atretic follicles (ATR), neither pyknotic granulosa cells nor hypertrophied follicular wall cells stain for MIS, whereas granulosa cells in a preantral follicle (PAF) next to the atretic follicle stain sharply. (From Ueno S, Takahashi M, Manganaro TF, Ragin RC, Donahoe PK, Cellular localization of mullerian inhibiting substance in the developing rat ovary, Endocrinology 1989;124(2):1000–6, with permission.)

proteins surrounding 7–8 cysteine residues. Interestingly, the 5'-flanking region of the gene does not contain any known consensus promoter/enhancer sequences, but work is underway to characterize specific nuclear regulatory proteins shown to bind to unique palindromic or dyad symmetry elements and to ascertain what differences exist in these factors that could explain the sexually dimorphic production of MIS (unpublished observations).

The human gene has been transfected into Chinese hamster ovary cells and, with methotrexate amplification of the construct, coupled to an SV40 early promoter. These cells produce 2–5 µg/mL of MIS in 4L Techne bioreactors (29) and provide the laboratory with a steady supply of recombinant human MIS (rhMIS) purified in the same manner as the bovine material.

Recent studies on the rhMIS demonstrated that, like TGFβ, MIS must be proteolytically cleaved to be biologically active. At least one of these cleavage sites is a plasmin-sensitive monobasic sequence at residue 427 (30), and at least one other cleavage site, N-terminal to this locus but as yet not clearly delineated, also exists (unpublished observations). Unlike the TGFβs, however, in which latent factor is activated by separation of the carboxy (C) and amino (N) terminal domains, MIS appears to require both fragments, as neither exhibits the ability to cause mullerian duct regression in the standard bioassay when tested alone. Recombi-

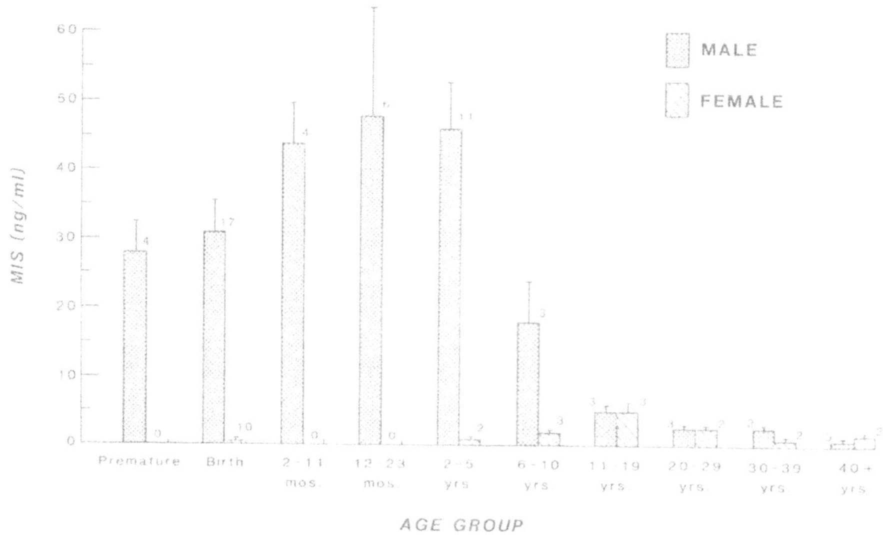

Fig. 5. Ontogeny of serum MIS levels in sera of human male and female subjects. Samples were collected from patients at the age ranges indicated. MIS levels were quantitated by ELISA. Numbers over the bars are the numbers of samples per group, and the brackets denote the standard error. (From Hudson PL, Douglas I, Donahoe PK, et al., An immunoassay to detect human mullerian inhibiting substance in males and females during normal development, J Clin Endocrinol Metab 1990;70:16–22, with permission.)

nation of the isolated N- and C-regions, on the other hand, restores much of the bioactivity in this system. Examination of rhMIS by protein staining in polyacrylamide gels under disulfide bond-reducing conditions shows that ~10% of the protein purified by immunoaffinity chromatography is already partially processed as it is secreted from the CHO cells (Fig. 6). Since processing of MIS in vivo may be a method by which the action of this growth regulator is modulated, we are currently evaluating its posttranslational modification to determine if target tissues play a role in this activity, or whether it is the function of the cells synthesizing MIS, or alternatively, whether MIS itself possesses the ability to modify its own activity.

Another area of great interest to our group is the role that N- and O-linked carbohydrate side chains of MIS play in MIS action. At present, we have no concrete evidence on this subject, but it is likely, as is the case with other serum-borne biological response modifiers, that the carbohydrate is required for the optimal effects of MIS to be manifest.

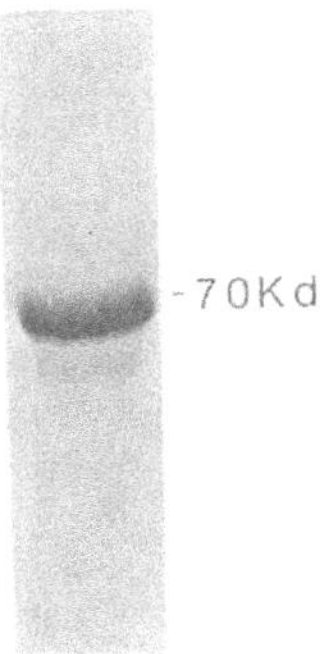

Fig. 6. Coomassie blue-stained immunoaffinity-purified human recombinant MIS run on 7.5% polyacrylamide gel electrophoresis under reducing conditions. Clearly demonstrated are the 70-kD monomeric form of MIS and the small amount of the 57-kD amino terminal fragment predicted by the proteolytic cleavage of rhMIS at residue 427. (From Pepinsky RB, Sinclair LK, Chow EP, et al., Proteolytic processing of mullerian inhibiting substance produces a transforming growth factor-β-like fragment, J Biol Chem 1988;263(35):18961–4, with permission.)

MIS AND THE REPRODUCTIVE TRACT

As the name originally given to the substance implies, the best-known role of MIS is to cause regression of the mullerian ducts in male embryos. This activity begins as the primitive gonad matures into a testis, and Sertoli cell function is realized. In the rat, this process begins at about day 14 of gestation, as evidenced by the appearance of bioactive MIS in the standard in vitro urogenital ridge assay (see Fig. 2).

It takes approximately 3 days to completely regress the mullerian duct in this assay system, with a minimum time of exposure of the ducts ranging from 24–36 h (21). Removing MIS from the ducts earlier results in only partial dissolution of the duct system. A body of evidence supports observations that the timing of MIS production coincides with the maximal sensitivity of the duct to MIS action. If one attempts to regress mullerian structures harvested from female rat embryos later in development than day 16, no activity is seen. These results indicate the possibility that a specific MIS receptor is only expressed transiently during development, thus protecting the mature female reproductive tract from the MIS produced after birth. These observations are supported by the clinical observations in adult human females in which high levels of MIS are found due to synthesis of the factor by granulosa or sex cord tumors (23). In the cases evaluated thus far, the uterus, cervix, and fallopian tubes appear grossly normal. Tumors of mullerian origin appear to be different, however, since a significant proportion of certain endometrial, ovarian, and fallopian tube neoplasias are growth inhibited by MIS in vitro (31).

The process of mullerian duct regression is characterized by several hallmarks observed using histological and immunohistochemical techniques (32). Early steps in regression include the clearing and condensation of the mesenchyme around the basement membrane followed by loss of the basement membrane entirely. Epithelial cells then become pyknotic, and the ducts begin to shrink in size and the duct lumen is lost. As other epithelial cells and connective tissue invade the area, the mullerian duct is no longer detectable, indicating complete regression. Coincident with these changes, elements of extracellular matrix, including fibronectin, collagen IV, laminin, and heparin sulfate, diminish and eventually disappear in the mullerian duct. Dose-response studies indicate that complete regression of mullerian ducts in vitro requires 1–2 μg/mL of MIS (or 7–14 nM). MIS can be inhibited in this in vitro bioassay by 50-μM vanadate, zinc at levels exceeding 400 μM, by epidermal growth factor (EGF) at approximately 1×10^{-7}M and by millimole levels of adenine and guanine nucleotides (11, 33–34). EDTA (0.5 mM) and fluoride (1.0 mM) were the only reagents tested that were shown to mimic MIS activity in vitro; however, the histologic pattern of duct regression varied significantly from that seen with MIS itself.

These observations lead us to speculate that the mechanism of action of MIS may indeed center about controlling intracellular phosphorylation events, as agents that increase phosphorylation state (such as EGF, zinc, and vanadate) block MIS, whereas compounds decreasing this process (EDTA and fluoride) had the opposite effect (11, 13–14, 34–36). The specific mechanisms remain to be established for these findings, as are those relating to the inhibition of MIS action by estrogens and potentiation by androgens and progestins described in mammalian and avian species (35, 37). It appears true, however, that proper male genital phenotype results from the balance of MIS activity and that of its natural inhibitors. Deviations from the normal programmed delivery of these factors in vivo would predictably result in a variety of defects, including retained mullerian ducts and other types of congenital reproductive tract anomalies.

MIS activity in the female is markedly different than that observed in males. MIS is not produced in the female until after birth. Such an occurrence would have disastrous effects on the developing female embryo by causing elimination of the reproductive tract. In fact, nature provides some evidence of what can happen in such a case. The Freemartin calf, a female calf with signs of virilization, particularly of the gonads, results from twin pregnancies in which a male and female embryo share a common placenta. The female is effected by the MIS produced by the fetal testes of her fraternal twin. More recently Josso and colleagues have shown that MIS incubated with fetal rat ovaries results in the formation of seminiferous tubule-like structures in the gonad and reduced estrogen production as the result of suppression of aromatase activity (38).

In the female, MIS is produced by the granulosa cells and we have provided evidence that a primary role for ovarian MIS may be to inhibit oocyte meiosis (9–12). If one follows the ontogeny of MIS in the female, it appears at puberty just at the time when ovarian cycles begin and oocytes are recruited to develop for ovulation. Prior to this time, the considerable meiotic division that occurs ceases at approximately the time when MIS synthesis begins. During estrus cycles in the rat, MIS levels, as measured by immunohistochemical means, increase until several hours before the evening of ovulation whereupon MIS production ceases just at the time of the Lutenizing hormone surge (24). Coincident with these events the oocyte undergoes meiosis, as evidenced by the disappearance of the germinal vesicle. A series of in vitro studies have shown that it is the drop in MIS that initiates the onset of meiosis since incubation of oocytes with rhMIS prevents the spontaneous germinal vesicle breakdown that occurs upon isolating oocytes into culture media (Fig. 7).

Interestingly, this effect is blocked by EGF, just as MIS-induced mullerian duct regression is inhibited by EGF (12) in the standard organ culture assays (11).

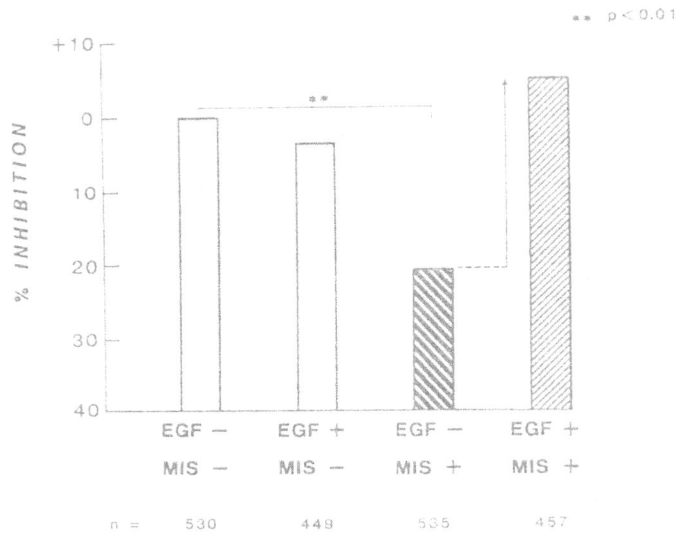

Fig. 7. Antagonistic effect of EGF on MIS inhibition of spontaneous rat oocyte meiosis, as measured by the percentage oocytes undergoing germinal vesicle breakdown ([number of oocyte without germinal vesicle]/[total number of oocytes] × 100). Five MIS samples purified from serum-free conditioned medium were used. Protein concentrations ranged from 23 to 520 ug/mL, and MIS was 1–20% pure. Krebs ringers bicarbonate (KRB) buffer with NP-40 at 10^{-7}g/mL containing the MIS samples inhibited oocyte meiosis significantly (20% inhibition compared to KRB medium with NP-40 [62% vs. 78%]; $P < 0.01$]). By adding EGF at 25 ng/mL to the incubation medium containing MIS, the inhibitory effect was antagonized (5% enhancement [79% vs. 75%]). KRB medium containing EGF at 25 ng/mL had no significant effect on rat oocyte meiosis (n = numbers of oocytes examined). (From Ueno S, Manganaro TF, Donahoe PK, Human recombinant mullerian inhibiting substance inhibition of rat oocyte meiosis is reversed by epidermal growth factor in vitro, Endocrinology 1988;123(3):1652–9, with permission.)

The role of MIS in the ovary, therefore, may be to prevent premature meiotic divisions in the developing oocyte, allowing it to properly mature for ovulation and subsequent fertilization. Here again, alterations in the normal ovarian follicular fluid milieu with respect to MIS levels and those of numerous other fluid proteins and factors may result in impaired fertility in the affected individual. It is not yet known whether MIS as produced by the granulosa cells has any function in the female beyond those just described, nor whether a similar role is played by MIS on developing sperm in the adult male. These latter topics are the subject of ongoing research in our laboratory.

MIS AND FETAL LUNG DEVELOPMENT

At present, the only other example of extra-Mullerian activities of MIS occurs in the male fetus and involves the developing lung (10). The reasons for considering the lung as a target tissue for MIS stem from the long-established fact that respiratory distress syndrome (RDS) is more prevalent in male infants than in females and that the other obvious male-specific hormone, testosterone, which has been shown to directly reduce lung development, acts in concert with MIS to cause mullerian duct regression. MIS seemed a likely candidate to explain, along with testosterone, the observed sexually dimorphic pattern of fetal lung development. Accordingly, a series of in vitro studies were begun in which the effects of MIS on fetal lung surfactant accumulation, in the form of disaturated phosphatidylcholine (DSPC) production, were assessed. Fetal day 17–18.5 female rat lungs incubated with ovarian fragments produced more DSPC than when incubated with testicular fragments. Furthermore, incubation of female fetal lungs with MIS reduced the level of DSPC production intermediate to those observed in the presence of fetal testis (Fig. 8).

This experimental result has been recapitulated in vivo by injecting MIS directly into developing rat fetuses at day 19.5 of gestation (39). In these studies MIS reduces DSPC formation in female lung tissue relative to vehicle-injected female litter mates (39).

It is of interest to note that in addition to corticosteroids and thyroid hormone, EGF and estrogen stimulate surfactant production, while androgens decrease their formation (10). Since EGF (and estrogen) are naturally occurring antagonists of MIS action on the urogenital ridge, the fetal lung represents another example of the MIS-EGF antagonist activity similar to what is observed for the estrogens and progestins as a naturally occurring pair of hormones with opposing activities.

MOLECULAR MECHANISM OF ACTION

The most productive clues as to the mechanism of action of MIS came from control experiments performed using the in vitro bioassay for MIS. In these studies it was determined that of several growth factor inhibitors tested, including nerve growth factor, insulin, TGFβ, fibroblast growth factor, EGF, corticosterone, and estradiol, only EGF and estradiol significantly blocked MIS activity (3, 11). Androgens and

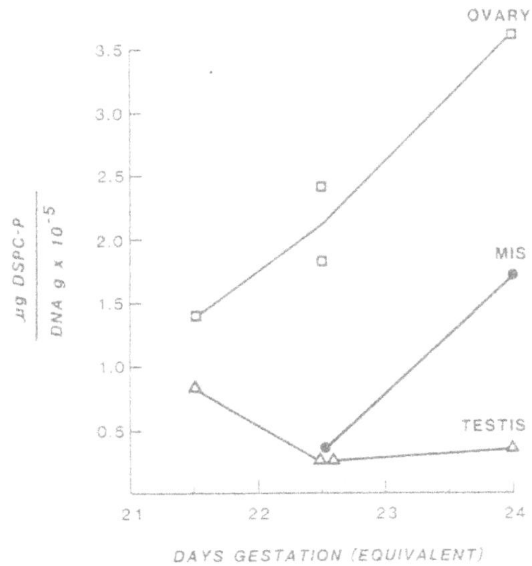

Fig. 8. Effects of fetal gonads on MIS incubated with minced female fetal rat lungs on disaturated phosphatidylcholine accumulation in organ culture. All viable female fetuses from 31 litters were put to death to generate 185 lung cultures. Each point represents 12–25 lung cultures pooled (each culture consists of twenty 1-mm^3 fragments from 17.0- to 18.5-day female fetal rat lung). Lung cultures incubated with either testis or MIS had disaturated phosphatidylcholine/DNA values that were significantly less than those incubated with ovary on day 22.5 ($P = 0.012$). (From Catlin EZ, Manganaro TF, Donahoe PK, Mullerian inhibiting substance depresses accumulation in vitro of disaturated phosphatidylcholine in fetal rat lung, Am J Obstet Gynecol 1988;159:1299–303, with permission.) Longer periods of exogenous MIS exposure lowers the DSPC level even further (data not shown).

progestins, on the other hand, augment MIS bioactivity, but do not produce regression on their own (35). Controls done to test constituents of buffers used in the purification of MIS revealed that of all components examined, only EDTA produced an MIS-like effect that could be reversed with zinc. High concentration of adenine and guanine nucleotides however, inhibited MIS, as did vanadate (34, 36).

The most productive of these observations proved to be those with EGF and vanadate. Both of these materials alter the phosphorylation state of the cell. EGF, after inducing autophosphorylation of its own receptor at specific tyrosine residues, phosphorylates intracellular proteins. Vanadate, on the other hand, blocks protein phosphatase activity and thereby indirectly favors the generation of phosphorylated proteins. Therefore, under conditions of increased cellular phosphorylation due to EGF or vanadate, MIS action was blocked. We reasoned that MIS may be acting in the opposite way; that is, to ultimately reduce the phosphorylation due to EGF and its subsequent effects on cell growth. To test this hypothesis we investigated the effects of MIS on the growth of a cell line (A431) stimulated to grow in soft agar by EGF. MIS coincubation clearly blocked the EGF-induced colony formation of these cells (13). Since EGF activity is modulated by the tyrosine autophosphorylation of its own receptor, we next examined the ability of MIS to inhibit this process and, therefore, all subsequent phospho-protein-mediated EGF effects. MIS in isolated A431 plasma membranes purified in the absence of calcium and in the presence of EDTA, or in whole cells, can completely inhibit the EGF-dependent autophosphorylation of tyrosine residues in the 170-kD receptor (Fig. 9) (13–14).

This effect is reversed by prior incubation of MIS with specific MIS antibodies and is not altered by high concentrations of EGF, ATP, or manganese, the cofactor required for EGF activity (13–14). Finally, since MIS does not reduce the level of phosphorylation of prestimulated EGF receptor, we do not feel that MIS is a phosphatase activity (13–14). Although protein kinase C, a known down-regulator of EGF activity, is absent from the isolated membrane preparations, it is possible that this enzyme may play a role in this process in vivo. Our current working hypothesis is that MIS binds to its own receptor and by so doing alters the ability of the EGF receptor to activate its tyrosine activity. This may be the result of a decreased affinity for EGF, but not a decrease in the number of binding sites, as it has been shown that MIS has no effect on this metameter, nor an inability to bind ATP, thus preventing catalysis (13). Our laboratory is currently attempting to

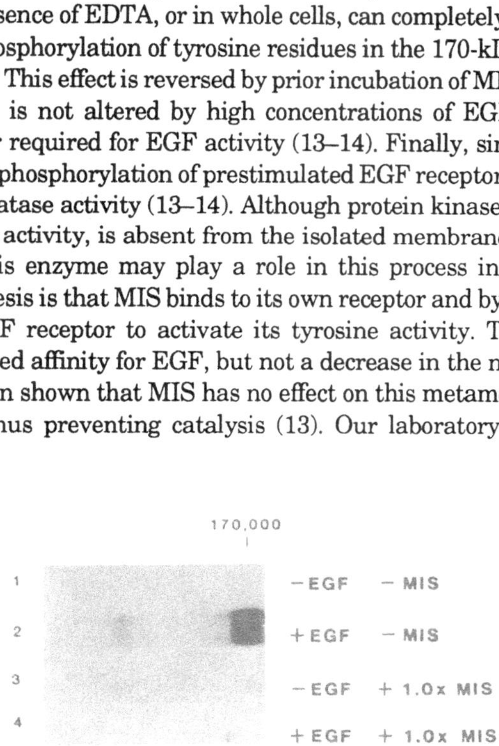

Fig. 9. Whole cell assay P_{TYR}Ab. The EGF receptor was immunoprecipitated from solubilized A431 cell extracts using antiphosphotyrosine antibody. In the absence of EGF and MIS (–EGF – MIS), there occurred no tyrosine phosphorylation at the EGF receptor. EG (+EGF – MIS) resulted in stimulated autophosphorylation of tyrosine residues at its receptor. Human recombinant MIS (+EGF + 1.0 × MIS), purified from serum-containing conditioned media, inhibited EGF-stimulated autophosphorylation of tyrosine residues at its receptor at a concentration of 50 nM. MIS alone (–EGF + 1.0 × MIS) caused no change in EGF receptor tyrosine phosphorylation compared with basal phosphorylation (–EGF – MIS). Autoradiogram: 3-h exposure. (From Cigarroa FG, Coughlin J, Donahoe P, White M, Uitvlugt N, MacLaughlin DT, Recombinant human mullerian inhibiting substance inhibits epidermal growth factor tyrosine kinases, Growth Factors 1989;1:179–91, with permission.)

characterize the putative MIS receptor in the A431 cell line as well as in fetal rat lung, as yet our most abundant source of naturally occurring receptor.

SUMMARY

Mullerian inhibiting substance, by virtue of its varied functions, now emerges as a growth-regulating protein that specifically inhibits EGF activities that are dependent upon tyrosine phosphorylation. In the fetus MIS action appears limited to the male, where it causes mullerian duct regression and attenuates lung development. In females, and perhaps also in males, MIS plays an important role in regulating germ cell meiotic activity. Whether other activities in the adult or fetal male exist is not yet fully understood. The discovery recently of a specific liver-derived inhibitor of insulin-dependent growth, pp63 (40), raises the possibility, when coupled to the MIS observations, that there exists a class of protein biologic response modifiers that regulate growth promoters by altering the catalytic activity of growth factor receptors. The fact that pp63 does not alter EGF activity (40) nor MIS insulin action speaks to the specificity of these proteins. It is interesting to note that although pp63 blocked insulin-dependent growth of a sensitive cell line, it did not inhibit other actions of insulin, such as glucose uptake, raising the possibility that such factors as pp63 and MIS will be useful in deciphering the effects of insulin and EGF, respectively, that are phosphorylation dependent from those independent of this receptor-mediated event.

Understanding the molecular mechanism of action of MIS as a member of the EGF-MIS antagonistic pair of factors will allow for greater insight into how the development of the male and female reproductive tract is regulated and perhaps identify a more logical manner to address the role MIS may play as a chemotherapeutic agent against tumors of mullerian origin.

REFERENCES

1. Palmer MS, Sinclair AH, Berta P, et al. Genetic evidence that ZFY is not the testis-determining factor. Nature 1989;342:937-9.
2. Koopman P, Gubbay J, Collington J, Lovell-Badge R. ZFY gene expression patterns are not compatible with a primary role in mouse sex determination. Nature 1989; 342:940-2.
3. Donahoe PK, Cate RL, MacLaughlin DT, et al. Mullerian inhibiting substance: gene structure and mechanism of action of a fetal regressor. Rec Prog Horm Res 1987;43:431-67.
4. Josso N, Picard J-Y, Tran D. A new testicular glycoprotein: Anti-mullerian hormone. In: Steinberger A, Steinberger E, eds. Testicular development, structure and function. New York: Raven Press, 1980:21-31.
5. Hutson J, Ikawa H, Donahoe PK. The ontogeny of mullerian inhibiting substance in the gonads of the chicken. J Pediatr Surg 1981;16:822-7.
6. Josso N, Vigier B, Tran D, Picard J-Y. Initiation of production of anti-mullerian hormone by the fetal gonad. Arch Anat Micr Exp 1985;74-96.
7. Takahashi M, Hayashi M, Manganaro TF, Donahoe PK. The ontogeny of mullerian

inhibiting substance in granulosa cells of the bovine ovarian follicle. Biol Reprod 1986;35:477-53.
8. Donahoe PK, Takahashi M, Ueno S, Manganaro TF. Mullerian inhibiting substance in the ovary. In: Hazeltine F, First N, eds. Meiosis inhibition molecular control of meiosis. New York: Alan R. Liss, 1988;153-76.
9. Takahashi M, Koide SS, Donahoe PK. Mullerian inhibiting substance as oocyte meiosis inhibitor. Mol Cell Endocrinol 1986;47:225-34.
10. Catlin EZ, Manganaro TF, Donahoe PK. Mullerian inhibiting substance depresses accumulation in vitro of disaturated phosphatidylcholine in fetal rat lung. Am J Obstet Gynecol 1988;159:1299-303.
11. Hutson JM, Fallat MW, Kamagata S, Donahoe PK, Budzik GP. Phosphorylation events during mullerian duct regression. Science 1984;223:586-89.
12. Ueno S, Manganaro TF, Donahoe PK. Human recombinant mullerian inhibiting substance inhibition of rat oocyte meiosis is reversed by epidermal growth factor in vitro. Endocrinology 1988;123(3):1652-9.
13. Coughlin JP, Donahoe PK, Budzik GP, MacLaughlin DT. Mullerian inhibiting substance blocks autophosphorylation of the EGF receptor by inhibiting tyrosine kinase. Mol Cell Endocrinol 1987;49:75-86.
14. Cigarroa FG, Coughlin J, Donahoe P, White M, Uitvlugt N, MacLaughlin DT. Recombinant human mullerian inhibiting substance inhibits epidermal growth factor tyrosine kinases. Growth Factors 1989;1:179-91.
15. Jost A. Sur la differenciation sexuelle de l'embryon de lapin remarques au sujet des certaines operations chirurgical sur l'embryon. CR Soc Biol 1946;140:461-2.
16. Jost A. Sur la differenciation sexuelle de l'embryon de lapin. Experiences de paraboise. CR Soc Biol 1946;140:463-4.
17. Jost A. Recherches sur la differenciation sexuelle de l'embryon de lapin. Arch Anat Micro Morph Exp 1947;36:271-315.
18. Jost A. Sur les derives mulleriens d'embryons de lapin des deux sexes castres a 21 jours. CR Soc Biol 1947;141:135-6.
19. Josso N, Picard JY, Tran D. The anti-mullerian hormone. Rec Prog Horm Res 1977;33:117-67.
20. Picon R. Action dy testicule foetal sur le development in vitro des canaux de Muller chez le rat. Arch Anat Micro Morph Exp 1969;58:1-19.
21. Donahoe PK, Ito Y, Marfatia S, Hendren WH. A graded organ culture assay for the detection of mullerian inhibiting substance. J Surg Res 1977;23:141-8.
22. Ueno S, Takahashi M, Manganaro TF, Ragin RC, Donahoe PK. Cellular localization of mullerian inhibiting substance in the developing rat ovary. Endocrinology. 1989; 124(2):1000-6.
23. Hudson PL, Dougas I, Donahoe PK, et al. An immunoassay to detect human mullerian inhibiting substance in males and females during normal development. J Clin Endocrinol Metab 1990;70:16-22.
24. Ueno S, Kuroda T, MacLaughlin DT, Ragin RC, Manganaro TF, Donahoe PK. Mullerian inhibiting substance in the adult rat ovary during various stages of the estrous cycle. Endocrinology 1989;125:1060-6.
25. Budzik GP, Powell SM, Kamagata S, Donahoe PK. Mullerian inhibiting substance fractionation by dye affinity chromatography. Cell 1983;34:307-14.
26. Shima H, Mudgett-Hunter M, Budzik GP, Kamagata S, Hudson PL, Donahoe PK.

Production of monoclonal antibodies for affinity purification of bovine mullerian inhibiting substance activity. Hybridoma 1984;3:201-14.
27. Budzik GP, Swann DA, Hayashi A, Donahoe PK. Enhanced purification of mullerian inhibiting substance by lectin affinity chromatography. Cell 1980;21:909-15.
28. Cate RL, Mattaliano RJ, Hession C, et al. Isolation of the bovine and human genes for mullerian inhibiting substance and expression of the human gene in animal cells. Cell 1986;45:685-98.
29. Epstein J, DesJardins EJ, Hudson PL, Donahoe PK. Stainless steel mesh supports high density cell growth and production of recombinant mullerian inhibiting substance. In Vitro Cell Dev Biol 1989;25(2):213-6.
30. Pepinsky RB, Sinclair LK, Chow EP, et al. Proteolytic processing of mullerian inhibiting substance produces a transforming growth factor-β-like fragment. J Biol Chem 1988;263(35):18961-4.
31. Fuller AF, Krane IM, Budzik GP, Donahoe PK. Mullerian inhibiting substance reduction of human cancers in a stem cell assay. Gyn Oncol 1985;22:135-48.
32. Donahoe PK, Budzik GP, Trelstad RL, Schwartz BR, Fallat ME, Hutson JM. Molecular dissection of mullerian duct regression. In: The role of extracellular matrix in development. New York: Alan R. Liss, 1984:573-95.
33. Donahoe PK, Hutson JM, Fallat ME, Kamagata S, Budzik GP. Mechanism of action of mullerian inhibiting substance. Annu Rev Physiol 1984;46:53-65.
34. Budzik GP, Hutson JM, Ikawa H, Donahoe PK. The role of zinc in mullerian duct regression. Endocrinol 1982;110:1521-5.
35. Ikawa H, Hutson JM, Budzik GP, MacLaughlin DT, Donahoe PK. Steroid enhancement of mullerian duct regression. J Pediatr Surg 1982;17:453-8.
36. Ikawa H, Hutson JM, Budzik GP, Donahoe PK. Cyclic adenosine 3', 5' monophosphate modulation of mullerian duct regression. Endocrinology 1984;114:1686-91.
37. Hutson JM, Ikawa H, Donahoe PK. Estrogen inhibition of mullerian inhibiting substance in the chick embryo. J Pediatr Surg 1982;17:953-9.
38. Vigier B, Forest M, Eychenne B, et al. Anti-mullerian hormone produces endocrine sex reversal of fetal ovaries. Proc Natl Acad Sci 1990;86:3684-8.
39. Catlin EA, Powell SM, Manganaro TF, et al. Sex-specific fetal lung development and mullerian inhibiting substance. Am Rev Respir Dis 1990;141:466-70.
40. Auberger P, Falquerho L, Contreves, et al. Characterization of a natural inhibitor of the insulin receptor tyrosine kinase-cDNA cloning and anti-mitogenic activity. Cell 1989;58:631-40.

12

Growth Factors Affecting Normal and Malignant Prostatic Cells

John T. Isaacs,[1,2] Ronald A. Morton,[2] Paula Martikainen,[1] and William B. Isaacs[2]

[1]The Johns Hopkins Oncology Center and The Brady Urologic Institute, Baltimore, Maryland, and [2]The Department of Urology, The Johns Hopkins School of Medicine, Baltimore, Maryland

The growth of any tissue, whether normal or malignant, depends upon the quantitative relationship between the rate of cell proliferation and cell death. In normal adult tissues, these rates are balanced such that a steady state (i.e., self-renewing) relationship is maintained in which the size of the normal tissue does not increase continuously with time. Studies in a variety of different tissues have demonstrated that this steady state balance is regulated by a series of both systemic and local growth factors. A fundamental characteristic of cancer is that unlike their normal counterparts, their rate of cell proliferation exceeds their rate of cell death, thus resulting in continuous net growth of the cancer. This does not mean, however, that cancer cells are unresponsive and therefore autonomous to such growth factors. On the contrary, many types of cancer cells respond to normal growth factors and often in a manner very similar to that of the normal cell of origin.

The fact that cancer cells continue to undergo net growth with time does demonstrate, however, that these cells have undergone change(s) that modifies their responsiveness to such growth factors. The identification of the specific change(s) within cancer cells that modifies their responsiveness to growth factors, allowing abnormal continuous growth, is thus an important area of research. In order to identify such change(s), an understanding of the growth factors requirements and responsiveness of the normal cell of origin for a particular cancer is necessary. The present chapter reviews the state of knowledge concerning the involvement of growth factors in the normal prostate. Based upon this review, a

Acknowledgments: The skillful help of Barbara Lee on the preparation of this manuscript is gratefully acknowledged. Studies reported in this paper were supported by HHS, NCI grant CA-15416.

comparison of the growth factor requirements and responsiveness will be made for malignant prostatic cells in an attempt to better define areas for future studies.

ANDROGEN: THE MAJOR PROSTATIC GROWTH FACTOR

Growth factors can be divided into two basic types (i.e., systemic vs. local) based upon the site of initial production of the factor. Systemic growth factors, more commonly termed hormones, are produced and secreted into the blood by specialized endocrine glands. Once in the blood, these hormones circulate systemically and can stimulate target cells throughout the body that possess high-affinity receptors. Such endocrine stimulation is well established. Recently, it has been demonstrated that besides systemic hormones, there is a series of local growth factors (e.g., epidermal growth factor [EGF], fibroblast growth factor [FGF], insulin-like growth factors I and II [IGF-I and IGF-II], transforming growth factors α and β [TGFα and TGFβ], etc.) that are involved as important regulators of cell number and function in a variety of tissues, including, possibly, the prostate. The ability of cells to respond to such local growth factors requires both the presence of high-affinity specific receptors for the particular growth factor and a sufficient concentration of the factor in the local microenvironment. Since the systemic concentration of these local growth factors is usually insufficient to elicit a biological response, the rate of local production of these growth factors is a major determinant as to whether or not a response is induced. The site of production of the local growth factors can be the same cell that responds to the factors (i.e., autocrine stimulation) or a neighboring cell (i.e., paracrine stimulation).

Androgen is the major systemic hormone for the prostate. The major source of androgen in the body is the testis, which synthesizes and secretes testosterone (i.e., the major physiological androgen) into the blood. The hypothalamic-pituitary-testis axis is tightly regulated so that the blood level of testosterone is maintained within a defined physiological range. Testosterone, derived from the blood, diffuses into prostatic cells. The prostatic cells contain 5α-reductase enzymatic activity that converts prostate testosterone into 5α-dihydrotestosterone (DHT) (1). Due to the efficiency of this prostatic 5α-reductase activity, the DHT concentration within the prostate is 4- to 5-fold higher than testosterone, and >95% of the androgenic steroid present within prostatic cell nuclei is DHT (2–3). In addition, the affinity of the androgen receptor for DHT is higher than for other endogeneously occurring steroids (4). DHT is thus believed to be the major intracellular effector of androgen action within the prostate (5).

NORMAL RESPONSE OF THE PROSTATE TO ANDROGEN

In order to understand the response of the prostate to androgen, an appreciation of the cellular organization of the gland is required. The prostate is a tubulo-alveolar gland composed of multiple secretory acini that are lined by epithelial cells. These acini drain into a system of branching epithelial ducts and tubules that eventually end in the prostatic urethra. The epithelial compartment is composed of two types of cells, a basal epithelial and a glandular epithelial cell. The basal epithelial cells are numerically less frequent than the glandular cells and are not as extensively developed morphologically for a secretory function (6–8). The function of the basal epithelial cells is not fully understood, but it has been suggested that a subset of basal cells may be the stem cells of the glandular epithelial cells (7). A well-developed basement membrane separates the epithelial acini and ducts from the surrounding fibromuscular stromal tissue. The prostatic stromal tissue is composed of fibroblasts, smooth muscle cells, endothelial cells, and assorted infiltrating cells (e.g., mast cells, lymphocytes, etc.). Thus, the prostatic basal and glandular cells receive nutrients and androgen that have traversed the endothelial cells, stromal cells, extracellular matrix, and the acinar basement membrane. Consequently, the stromal compartment has ample opportunity to modify the epithelial cell microenvironment.

The prostate normally undergoes two distinct phases during the lifetime of the host (9). For the human prostate, the first, or growth phase, begins around the age of 10 years and continues until approximately 20 years of age, when the human prostate reaches its normal adult size (i.e., ~20 g) (10). Between the ages of 10 and 20 years, the rate of prostatic growth is exponential, with a prostatic weight-doubling time of 2.78 years (11). This period of exponential growth corresponds to the time period when the serum testosterone levels are rising from their initial low levels seen before the age of 10 to the high levels seen in an adult male (12). If an individual is castrated before the age of 10, the serum testosterone levels do not rise to their normal adult level, and the proliferative growth of the human prostatic epithelial cells between 10 and 20 years of life is completely blocked (13). These results demonstrate that a physiological level of androgen is chronically required for the normal growth of the prostate.

After 20 years of age, the human prostate, having reached its maximum adult size, normally ceases continuous net growth, and the second, or maintenance, phase of the prostate begins (9). This does not mean, however, that the cells of the normal adult prostate are not continuously turning over with time, only that the rate of prostatic cell proliferation is balanced by an equal rate of prostatic cell death such that neither involution nor overgrowth of the gland normally occurs with time during the maintenance phase. The normal adult prostate during its maintenance phase is thus an example of a steady state, self-renewing tissue (9). If an adult male whose prostate is in this steady state-maintenance condition is castrated, the serum testosterone levels rapidly decrease to low values comparable to those seen in intact males younger than 10 years of age. As a result, the prostate rapidly involutes. This involution is due to a major loss in the epithelial, not the stromal, compartment of the prostate (7, 14, 15). This involution demonstrates that the normal prostatic epithelial compartment chronically requires a physiological level of androgen during its maintenance, as well as during its earlier growth phase.

ANDROGEN DEPENDENCY OF PROSTATIC GLANDULAR CELLS

The chronic requirement for androgen by the epithelial compartment is due to the fact that androgens regulate the total prostatic epithelial cell number by affecting both the rate of glandular cell proliferation and the rate of glandular cell death. Androgen does this by chronically stimulating the rate of cell proliferation (i.e., agonistic ability of androgen) while simultaneously inhibiting the rate of cell death (i.e., antagonistic ability of androgen) of the prostatic glandular epithelium (16). If a sufficient systemic androgen level is not chronically maintained (e.g., following castration), then prostatic glandular cells rapidly die via the activation of an energy-dependent cascade of biochemical and morphological change, collectively referred to as programmed cell death (15–24). The prostatic glandular cell is thus an example of a cell that is chronically androgen dependent. In contrast, prostatic basal epithelial cells do not activate this programmed-cell-death pathway following androgen ablation. Thus, the prostatic basal cells are not androgen dependent (7, 15).

The agonistic ability of androgen to stimulate prostatic glandular cell proliferation in vivo is well established. For example, if an animal or human is castrated and the prostate allowed to involute and then the host is given sufficient exogenous androgen replacement, the prostatic glandular epithelial cells are recruited out of G_0 and into the cell cycle (25). This recruitment results in a proliferative response that restores the prostatic glandular epithelial cell number (26). While the agonistic ability of androgen on prostatic glandular cell proliferation has been well established by a variety of studies, the additional ability of androgen to antagonistically inhibit the rate of death of these glandular cells has only recently been fully appreciated (16).

The importance of the androgenic regulation of prostatic glandular cell death is illustrated by the following observations. In the adult male rat, only 2% of the total prostatic cells die per day when the serum testosterone level is sufficient for the chronic steady state maintenance of the gland; that is, the cell proliferation rate per day is also 2% to balance this normal rate of cell death (16). Within 12 h after castration of an adult male rat, serum testosterone decreases to below 2% of the value present in an intact host (18). This rapid decline in serum testosterone results in the prostatic DHT concentration decreasing within the first 24 h following castration to below a critical threshold value that induces the eventual death of the androgen-dependent prostatic glandular epithelial cells (18, 27). By 2 days following castration, the percentage of total prostatic glandular cells dying per day increased nearly 10-fold from the value in intact hosts to a value of ~20% per day (16, 18). This death of the prostatic glandular epithelial cells occurs over a 7- to 10-day period following androgen ablation and is an active, energy-dependent process involving a cascade of biochemical and morphological changes, collectively referred to as programmed cell death (15–24). Associated with this programmed cell death is the enhanced expression of a series of genes within the prostate. These include an increase in the expression of the $TGF\beta_1$ gene (20), c-*myc* proto-oncogene (21), c-*fos* proto-oncogene (21), heat shock 70-Kd gene (21), testosterone-repressed prostatic message 2 gene (22), glutathione S-transferase Xb_1 gene (28), and a series of proteins of unknown function (29).

The role of these epigenetic changes in the programmed cell death of prostatic glandular cells is presently unknown. It is known, however, that like other systems in which programmed cell death occurs (30–33), prostatic cell death induced by androgen ablation initially involves an increase in the free Ca^{++} concentration that initiates fragmentation of genomic DNA (15, 18, 19). This DNA fragmentation occurs via activation of a Ca^{++}-Mg^{++}-dependent endonuclease(s) within the prostatic glandular cell nucleus that degrades the genomic DNA into nucleosomal oligomers (i.e., multiples of a 180-base-pair subunit) lacking intranucleosomal breaks in the DNA (18–19). This DNA fragmentation is subsequently followed by irreversible morphological changes, termed apoptosis (34), that characteristically involve chromatic condensation, nuclear disintegration, cell surface bleeding, and, eventually, cellular fragmentation into a cluster of membrane-bound apoptotic bodies within the prostate (18, 19, 35). Comparisons of the temporal induction of DNA fragmentation, appearance of apoptotic bodies, and the loss of prostatic glandular cell number following castration have demonstrated that the fragmentation of genomic DNA is the irreversible commitment step in the death of the androgen-dependent prostate glandular cells and does not occur as a result of the cells already being dead (15).

MECHANISM OF ACTION OF ANDROGEN WITHIN THE PROSTATE

Due to its agonistic ability on prostatic glandular cell proliferation, coupled with its antagonistic ability on prostatic glandular cell death, androgen is the major systemic growth factor for the prostate. Within the prostate of an adult host, testosterone is converted via 5α-reductase to DHT (36), which binds to high-affinity androgen receptors present within the nuclei of prostatic epithelial cells (4–5). These receptors have a higher affinity for DHT than for testosterone or any other endogeneous steroid (4–5). It is believed that it is the binding of DHT with these nuclear androgen receptors that initiates the action of androgen within the adult prostate. It has been demonstrated that a critical threshold concentration of intracellular DHT is required in order to stimulate prostatic epithelial cell proliferation while simultaneously inhibiting prostatic epithelial cell death (27).

Presently, there is a series of major, unresolved questions concerning the mechanism of action of androgen within the prostate. The first question is whether androgen acts directly on the prostatic glandular cells themselves without the requirement for the cooperation of the prostatic stromal cells. If prostatic stromal cells are required, this would strongly suggest that paracrine growth factors (e.g., EGF, FGF, IGF, etc.) are produced by the prostatic stromal cells in response to androgen and that these paracrine growth factors, not androgen, regulate the prostatic glandular cell number (i.e., glandular cells under paracrine regulation). In contrast, if prostatic stromal cells are not required for prostatic glandular cell response to androgen, then the question arises as to whether androgen induces the glandular cells themselves to produce local prostatic growth factors (i.e., autocrine regulation) or whether androgen itself mediates all of the response directly (i.e., endocrine regulation). In order to formulate experimental approaches to resolve the endocrine versus paracrine versus autocrine question, a review of the types of growth factors and their receptor present within normal and neoplastic prostate tissue is required.

GROWTH FACTORS AND THEIR RECEPTORS IN NORMAL AND NEOPLASTIC PROSTATIC TISSUE

Both normal and neoplastic prostatic tissue (i.e., benign prostatic hyperplasia [BPH] as well as malignant prostatic cancers) possess a variety of local growth factors and high-affinity receptors for these local growth factors. For example, epidermal growth factor (EGF) has been isolated from human (37), guinea pig (38), and rat normal prostatic tissue (39–40). Evidence has been presented that prostatic EGF-related mitogen (PEM) is a major growth factor present in the normal rat prostate (39–40) and that EGF-like factors can be secreted into the rat prostatic urethra (41). EGF has been identified in both human BPH and prostatic cancer tissue (42). In addition, EGF levels are very high in human prostatic fluid (43). EGF-like growth factors and TGFα (i.e., a factor that also binds to the EGF receptor) are made and secreted into the culture media in vitro by certain human prostatic cancer cell lines (44–45). EGF receptors are present on the cell membranes of normal rat prostatic cells (46), rat prostatic cancers (47), normal human prostatic tissue, human BPH, and human prostatic cancers (47–50). In human BPH tissue, EGF receptors are localized on the basal epithelial cells and not on the glandular epithelial or stromal cells (49). In both normal rat prostate (46) and human BPH tissue (50), EGF receptors are down-regulated by androgen. In contrast, EGF receptors are up-regulated by androgen in certain human prostatic cancer cell lines in vitro (51).

Platelet-derived growth factors (PDGF) I and II are synthesized and secreted into the media of certain androgen-independent human prostatic cancer cell lines in vitro (52). PDGF receptors, however were not detected in these same human prostatic cancer cell lines (52). Insulin-like growth factor (IGF) II is synthesized and secreted into the media of certain androgen independent rat prostatic cancer cell lines in vitro (53).

Fibroblast growth factors (i.e., acidic FGF [aFGF] and basic FGF [bFGF]) have been isolated from normal rat prostatic tissue (39, 54), rat prostatic cancer tissue (54), normal human prostatic tissue (55), human BPH (55–60), and human prostatic cancer tissue (54–60). The total FGF content in BPH tissue has been reported to be 2- to 4-fold higher than in normal or cancerous prostates (55). Normal rat ventral prostatic tissue expresses mRNA for aFGF in an age-dependent manner (i.e., expresses a significant decrease of aFGF mRNA after 14 weeks of age, a time when the ventral prostate slows its normal growth rate) with only a small amount of expression of the bFGF mRNA (61). Basic FGF is secreted, however, by rat prostatic tissue into the seminal fluid (62). Androgen-responsive rat prostatic cancers express aFGF mRNA, while androgen-independent, highly metastatic rat prostatic cancers express both aFGF and bFGF mRNA (61). Normal rat prostatic epithelial cells and stromal cells from rat prostatic cancer express only the aFGF mRNA when in in vitro cell culture (61). Normal rat prostatic epithelial cells and rat prostatic cancer cells possess high-affinity FGF-binding sites (61) (i.e., both aFGF and bFGF bind to the same FGF receptor [63]). In contrast to these rat studies, normal human prostatic tissue, human BPH tissue, and human prostatic cancer tissue express bFGF mRNA, but not aFGF mRNA (64). In addition, the expression of bFGF mRNA is elevated in human BPH tissue as compared to the normal prostate (65).

$TGF\beta_1$ mRNA is expressed in normal rat prostatic tissue (20). The level of $TGF\beta_1$ mRNA is down-regulated by androgen (20). Rat prostatic cancer cells also synthesize and secrete TGFβ when maintained in cell culture (66). $TGF\beta_1$ and $TGF\beta_2$ mRNA are also expressed by normal human prostatic tissue and by human BPH tissue (65); the level of $TGF\beta_2$ mRNA, but not $TGF\beta_1$ mRNA, appears to be increased in BPH as compared to normal prostatic tissue (65). The receptor for TGFβ has been identified in normal rat prostatic tissue where its level is negatively regulated by androgen (67).

EVIDENCE OF THE PARACRINE REGULATION OF PROSTATIC EPITHELIAL CELLS

Cunha and his associates have demonstrated the critical importance of prostatic stroma in the embryonic and neonatal growth of the prostatic epithelial cells (68). By recombining murine epithelium and mesenchyme from the embryonic urogenital sinus (i.e., prostatic anlage) of normal and testicular feminized mice, Cunha has demonstrated (*a*) that the morphogenesis and growth of the prostatic epithelium during the embryonic and neonatal period is dependent upon androgen; (*b*) that at these early stages of growth, androgen receptors are present only in the prostatic stromal, not the epithelial, cells of the developing prostate; and (*c*) the normal androgen-dependent growth of the prostatic epithelium can be induced when androgen receptor-positive wild-type urogenital sinus stromal cells are recombined with prostatic epithelium derived from testicular feminized mice that are genetically deficient in androgen receptors. These results strongly suggest that the growth of the prostatic epithelium during embryonic and neonatal periods is regulated by local paracrine growth factors produced by the prostatic stromal cells in an androgen-dependent manner.

While the androgen-dependent regulation of prostatic epithelial growth via the prostatic stroma is well established during early development, the effects of prostatic stroma on the prostatic epithelial cells in the adult prostate have not been resolved. McKeehan and associates have demonstrated that adult rat ventral prostatic epithelial cells can be grown in a serum-free media not containing any androgen if EGF, FGF, and the systemic growth factors insulin and glucocorticoid are included in the serum-free media, which also contains cholera toxin to elevate the intracellular levels of cyclic-AMP (69). Addition of androgen to the serum-free media had no effect on prostatic epithelial cell number. The inability of androgen to stimulate prostatic epithelial cells cultured in serum-free media containing EGF, FGF, insulin, glucocorticoid, and cholera toxin, has also been reported by Nishi, et al. (70) for normal rat dorsolateral prostatic epithelial cells and by Peehl and

Stamey (71) and Chaproniere and McKeehan (72) for normal human, BPH, and malignant prostatic epithelial cells.

One possible explanation of these interesting findings is that as discussed earlier, only the prostatic glandular epithelial cells are androgen dependent. In contrast, the prostatic basal epithelial cells are not dependent upon continuous androgen for their maintenance in vivo (7, 15). A series of studies has demonstrated that when prostatic tissue is established in primary culture, it is the prostatic basal cells, not the prostatic glandular cells, that grow in vitro (73–76). Thus, it is possible that the reason why the previous in vitro studies have failed to demonstrate an effect of androgen on the prostatic epithelial cells in in vitro culture is that these cultures contained only prostatic basal epithelial cells. If this is correct, it would not be unexpected that androgen has little, if any, effect on these basal cell cultures. There are presently available a wide range of monoclonal and polyclonal antibodies to the epithelial-specific keratin family of intermediate filament proteins. It has been demonstrated that the prostatic glandular and basal cells express both common and unique keratin proteins (77). There are keratin antibodies that specifically react with only the prostatic basal, and not the glandular, epithelial cells (77). Thus, using these basal cell-specific keratin antibodies, the issue of whether these previously studied cells are of prostatic basal cell origin should be directly answerable.

If these cultures are shown to be not of prostatic basal cell origin, but instead are derived from the prostatic glandular cells, this would suggest that either (*a*) androgen induces the autocrine production of EGF, FGF, and other local growth factors by normal and neoplastic prostatic epithelial cells directly; or (*b*) androgen induces the paracrine production by the supporting stromal cells of EGF, FGF, and/or other local growth factors. In either case, however, the presence of exogenous EGF and FGF added to the in vitro culture media would abrogate the requirement for androgen to stimulate the growth of either normal or neoplastic prostatic epithelial cells in culture.

Kabablin, et al. have demonstrated that the clonal growth of human prostatic epithelial cells in in vitro culture is stimulated by soluble growth factors released by fibroblasts (78). This stimulation was demonstrated by coculturing human prostatic epithelial cells with fibroblasts. Epithelial growth in coculture with fibroblasts was greater than could be obtained in isolated culture in serum-free media containing exogenous EGF, FGF, insulin, glucocorticoid, and cholera toxin (78). Not only human prostatic, but also adult human skin, human fetal lung, and mouse 3T3 fibroblast were capable of this stimulation (78). The fibroblasts were able to compensate for the decrease in stimulatory effect in culture when the exogenously added EGF and insulin, but not FGF, were deleted from the serum-free media (78).

This last finding is rather unexpected since Story, et al. demonstrated that when human prostatic-derived fibroblasts were cultured in vitro, they produced and released FGF (79). A possible explanation for the inability of fibroblasts to compensate for the deletion of FGF in the culture media in the studies of Kabablin, et al. is that in these latter studies the fibroblasts remain viable for only 20–40 h in in vitro coculture within the human prostatic epithelial cells (78). This possibility is also supported by the fact that in the studies of Kabablin, et al., formalin-fixed fibroblast monolayers and extracellular matrix prepared from the fibroblast cultures failed to stimulate human prostatic epithelial growth (78). In addition, growth factors, such as aFGF, bFGF, PDGF, EGF, and the like, were not able to stimulate these human prostatic epithelial cultures when given as single growth factors alone (78). This suggests that several growth factors, rather than a single one, are required for human prostatic epithelial cell growth.

Along these lines, McKeehan and associates demonstrated that normal rat prostatic epithelial cells require both EGF and FGF for maximal growth in serum-free cell culture (80). In additional studies it was demonstrated that exogenously added $TGF\beta_1$ was able to inhibit the normal rat prostatic epithelial cell growth induced in serum-free culture containing EGF plus FGF, no matter what the concentration of the latter two growth factors (81). In contrast, rat prostatic cancer cells require either EGF or FGF for maximal growth (80). Like normal rat prostatic epithelial cells, rat prostatic cancer cells were growth-inhibited by adding $TGF\beta_1$ to serum-free media containing EGF alone. Increasing the concentration of EGF in the media could not overcome this $TGF\beta_1$ inhibition (81). The dose-response requirement for rat prostatic cancer cells maintained in serum-free media containing FGF was likewise affected by adding $TGF\beta_1$ to the media; however, if the FGF concentration was increased to a high enough level, the $TGF\beta_1$ inhibition of growth could be overcome (81).

IN VIVO RESPONSE OF NORMAL PROSTATIC GLANDULAR CELLS TO GROWTH FACTOR ANTAGONISTS

As reviewed, there is a large body of circumstantial evidence implicating nonandrogen growth factors as important regulations of both normal and neoplastic prostatic cell number. In an attempt to develop more direct proof of the involvement of nonandrogen growth factors in regulation of prostatic cell number, known growth factor antagonists were tested for their effects on the growth of prostatic glandular cells. To do this, advantage can be taken of the ability to reproducibly induce prostatic glandular cell proliferation in the involuted ventral prostate of previously castrated male rats following exogenous androgen replacement. In this model, male rats are castrated, and the animals are untreated for 10 days to allow the loss (i.e., death) of >80% of the ventral prostatic glandular epithelial cells (7, 15). Animals are then given exogenous androgen replacement to restore the serum testosterone level to a value equal to that seen in a noncastrated adult male rat (i.e., 1- to 2-ng testosterone/mL serum). Such androgen treatment results in the proliferative regrowth of the prostatic glandular epithelial cells such that the number of these cells is fully restored by 2 weeks of treatment (7, 14, 26). In addition, such treatment results in a ~10-fold increase in the bFGF mRNA levels within 1 day following exogenous androgen replacement (82). This suggests that bFGF is an excellent candidate as a paracrine/autocrine intermediate in the androgen-induced regrowth of the rat ventral prostate. Thus, castrated male rats can be used as an in vivo model system to test the ability of various growth factor antagonists, particularly those targeted at FGF, to block the androgen-induced proliferative regrowth of the ventral prostatic glandular cells.

Suramin is a polysulfonated naphthylurea similar in structure to the diazo dyes trypan red and trypan blue. Properties of the drug include the ability to block the action of a number of growth factors, presumably by inhibiting their ability to bind to their receptors. This has been shown for PDGF, TGFβ, EGF, and bFGF (83–85). For example, both the mitogenic and motility-stimulating activities of PDGF and FGF on endothelial cells can be eliminated by suramin (85). Protamine sulfate, a low-molecular-weight, strongly basic protein used in the reversal of heparin anticoagulation, exhibits many of the same properties as suramin (85). Unlike suramin, however, protamine sulfate is a more specific antagonist for bFGF and PDGF (86–87).

To determine if growth factor antagonists could inhibit androgen-induced regrowth of the normal rat ventral prostate, suramin and protamine sulfate were administered to orchiectomized rats concurrent with androgen replacement. Because considerable evidence suggests that the growth of prostatic epithelium might be particularly sensitive to growth factor inhibition, the National Cancer Institute has entered a number of patients with stage D2 prostate cancer into phase II trials with suramin (88). In these clinical trials, serum suramin levels greater than 200 μg/mL have shown efficacy against advanced prostate cancer (88). For the rat studies, a number of intravenous (IV) doses of suramin were tested, and it was demonstrated that 300-mg/kg IV followed by 100-mg/kg IV on day 5 was the maximal dose tolerated with acceptable mortality. At this dose schedule, 20% of rats receiving suramin died prior to sacrifice. Using the 300-mg/kg loading dose followed by 100 mg/kg on day 5, the daily mean suramin serum levels were above 200 μg/mL for ~10 days. The dose of protamine sulfate used in our studies was 120 mg/kg given subcutaneously twice a day. This dose was chosen since previous studies by Taylor and Folkman (89) demonstrate that it could inhibit tumor growth in rats. This dose of protamine sulfate was also shown to be the maximally tolerated dose without inducing a significant morbidity and mortality (89).

Using the dose schedule outlined, neither suramin nor protamine sulfate had any effect on the androgen-induced regrowth of the normal rat ventral prostate. There was no difference in ventral prostate wet weight or DNA content between castrate animals receiving testosterone and castrates receiving testosterone plus suramin or protamine sulfate.

The studies of McKeehan, et al. (81) demonstrated that $TGF\beta_1$ could inhibit the in vitro growth of the rat ventral prostatic epithelial cells induced in serum-free media containing EGF, FGF, insulin, glucocorticoid, and cholera toxin. In addition, $TGF\beta_1$ mRNA levels increase during the time period when ventral prostatic glandular cell proliferation decreases in vivo following androgen ablation (20). These results suggest that $TGF\beta_1$ may be an important in vivo negative regulator of ventral prostatic glandular proliferation. To evaluate this possibility more directly, the ability of $TGF\beta_1$ to inhibit ventral prostatic glandular cell proliferation was tested using as a test system the previously described restoration of the involuted ventral prostate of castrated male rats induced by exogenous androgen replacement. In these experiments male rats were castrated, and 10 days were allowed for the ventral prostates to involute. Animals were then operated on to insert a catheter through the capsule of the ventral prostate. The other end of the catheter was connected to an Alza brand osmotic minipump that contained either vehicle alone or vehicle containing purified porcine TGFβ_1. The concentration of the TGFβ_1 in the pumps was formulated so that 40-ng TGFβ_1/day was directed to the ventral prostate. After a 1-day pretreatment period, the animals were started on exogenous testosterone replacement. The dose of exogenous testosterone replacement used restored the ventral prostatic wet weight and total prostatic DNA content to values equivalent to that of age-matched, noncastrated (i.e., intact) male rats by 2 weeks of treatment. In animals in which the ventral prostate was continuously exposed directly to 40 ng/day of TGFβ_1, there was no inhibition of the ventral prostatic wet weight or DNA regrowth.

IN VIVO RESPONSE OF MALIGNANT PROSTATE CELLS TO GROWTH FACTOR ANTAGONISTS

The cell of origin for most prostatic cancers is believed to be the prostatic glandular cell. This belief is based upon morphologic criteria (i.e., the majority of prostatic cancers are adenocarcinomas) and the observation that most prostatic cancers react positively with antibodies to cytokeratin isoforms present within glandular, not basal, prostatic epithelial cells (90–91). Since the majority of prostatic cancers appears to be derived from androgen-dependent prostatic glandular cells, it is not unexpected that the majority of men with prostatic cancers are responsive to androgen ablation therapy (92). Based upon the previously reviewed biochemical studies demonstrating the presence of growth factors (i.e., EGF, FGF, etc.) within prostatic cancer tissue and the ability of prostatic cancer cell lines to respond in vitro when such growth factors were added to the culture media, the ability of suramin and protamine sulfate to inhibit the in vivo growth of prostatic cancer was tested.

As a model, the serially transplantable Dunning R-3327 AT2 rat prostatic cancer was used. The AT2 is a rapid-growing, androgen-independent, prostatic cancer that is maintained in inbred Copenhagen (Cop) male rats (93). Cop male rats were inoculated with the AT2 tumor, and the tumor was allowed to grow to starting size of ~0.4 cm^3 before the animals were randomized into (*a*) a control group receiving no further treatment, (*b*) a group injected intravenously once with 300-mg suramin/kg body weight, or (*c*) a group injected twice daily with 120-mg protamine sulfate/kg body weight. The tumor size in each animal was individually measured with microcalipers over a 10-day period following initiation of treatment and the tumor doubling time determined as described previously (94). In the control untreated animals, the AT2 grew with a 2.4 ± 0.1 day doubling time. Treatment of animals with either suramin or protamine sulfate resulted in neither an involution of the AT2 tumor nor a cessation of continuous net growth. Such treatment did result, however, in a substantial decrease in AT2 growth rate (i.e., increase in tumor doubling time). For the suramin-treated animals, the growth rate increased to a doubling time of 3.3 ± 0.1 days; in the protamine sulfate-treated animals, the doubling time increased to a value of 3.1 ± 0.2 days. These changes in growth rate translated into the AT2 tumor size 10 days following treatment, being 4.2 ± 0.3 cm^3 in the suramin group and 3.6 ± 0.3 cm^3 in the protamine group versus 7.1 ± 0.9 cm^3 in control group. These results demonstrate that in contrast to their inability to inhibit the regrowth of normal prostatic glandular cells induced in castrated rats by androgen, suramin or protamine sulfate can inhibit the in vivo growth of androgen-independent rat prostatic cancer cells.

CONCLUSION

There is a large body of circumstantial evidence that growth factors, in addition to androgen, are involved in the regulation of both normal and neoplastic prostatic tissue cell number. There has been little in vivo data, however, to directly support the requirements for specific growth factors in the regulation of normal prostatic cell numbers. Preliminary in vivo data using the Dunning system of serially transplantable rat prostatic cancers has demonstrated that both suramin and protamine sulfate therapy can reduce the growth rate of even androgen-independent prostatic cancers. Whether these therapeutic responses are due to direct inhibitory effects on the prostatic cancer cells themselves or upon the angiogenesis that cancers must induce to grow (89) is unclear. This latter possibility is not trivial since both suramin (85) and protamine sulfate are known to have anti-angiogenic effects on endothelial cells (85, 89). If the response of the Dunning AT2 rat prostatic cancer to suramin or protamine sulfate is not due to an anti-angiogenic effect, then this suggests that prostatic cancer cells are more sensitive to anti-growth factor therapy than are the normal prostatic glandular cells. Due to its clinical relevance with regard to the use of growth factor antagonist therapy for human prostatic cancer, this important issue will have to be experimentally resolved. As part of these future studies, the possibility that growth factors are produced by bone cells that can stimulate human prostatic cancers needs to be resolved. Human prostatic cancer metastases are most commonly found in the bone, where their growth rate can be considerably faster than that of the primary cancer present within the prostate (95). Recently, novel mitogenic factors produced by bone marrow stromal cells have been identified that can stimulate the growth of human prostatic cell lines (96). Since the mitogens are not previously described growth factors, (i.e., not EGF, FGF, PDGF, TGFα, TGFβ, interleukin 1–4 or 6, GM-CSF, G-CSF, or M-CSF), there is a possibility that agents that could block either the production of these factors by the bone marrow stromal cells or antagonize their effect on prostatic cancer cells could be given systemically to prostatic cancer patients without the degree of toxicity induced by agents targeted at the more general growth factors, such as FGF.

REFERENCES

1. Wilson JD, Gloyna RE. The intracellular metabolism of testosterone in the accessory organs of reproduction. Recent Prog Horm Res 1970;26:309-36.

2. Siiteri PK, Wilson JD. Dihydrotestosterone in prostatic hyperthrophy, 1. The formation and content of dihydrotestosterone in the hypertropic prostate of man. J Clin Invest 1970;49:1737-45.
3. Bruchovsky N. Comparison of the metabolites found in rat prostate following the *in vivo* administration of seven natural androgens. Endocrinology 1971;89:1212-8.
4. Liao S, Fang S. Receptor-proteins for androgens and the mode of action of androgens on gene transcription in ventral prostate. Vitam Horm 1969;27:17-90.
5. Mainwaring WIP. The mechanism of action of androgen. In: Monographs in endocrinology. New York: Springer-Verlag, 1977.
6. Evans GS, Chandler JA. Cell proliferation studies in the rat prostate, I. The proliferative role of basal and secretory cells during normal growth. Prostate 1987;10: 163-78.
7. English HD, Santen RJ, Isaacs JT. Response of glandular versus basal rat ventral prostatic epithelial cells to androgen withdrawal and replacement. Prostate 1987;11: 229-42.
8. Evans GS, Chandler JA. Cell proliferation studies in the rat prostate, II. The effects of castration and androgen-induced regeneration upon basal and secretory cell proliferation. Prostate 1987;11:339-52.
9. Isaacs JT. Control of cell proliferation and cell death in the normal and neoplastic prostate: A stem cell model. In: Rodgers CH, Coffey DS, Cunha G, Grayhack JT, Hinman F, Horton R, eds. Benign prostatic hyperplasia; vol. 2. Bethesda, MD: NIH Pub. No. 87-2881, 1987:85-94.
10. Boyd E. Growth, including reproduction and morphological development. In: Altman, Dittmer, eds. Biological handbooks. Fed Am Soc Experimental Biology, Washington, D.C., 1962:346-8.
11. Isaacs JT. Common characteristics of human and canine benign prostatic hyperplasia. In: Kimball FA, Buhl AE, Carter DB, eds. New approaches to the study of benign prostatic hyperplasia. New York: Alan R. Liss, 1984:217-34.
12. Frasier SD, Gafford F, Horton R. Plasma androgens in childhood and adolescence. J Clin Endocrinol Metab 1969;29:1404-8.
13. Moore RA. Benign hypertrophy and carcinoma of the prostate. In: Twombly G, Packs G, eds. Endocrinology of neoplastic diseases. London: Oxford University Press, 1974:194-212.
14. DeKlerk DP, Heston WDW, Coffey DS. Studies on the role of macromolecular synthesis in the growth of the prostate. In: Grayback JT, Wilson JD, Scherbenske MJ, eds. Benign prostatic hyperplasia. DHEW Pub. No. (NIH) 76-1113, 1975:43-52.
15. English HF, Kyprianou N, Isaacs JT. Relationship between DNA fragmentation and apoptosis in the programmed cell death in the rat prostate following castration. Prostate 1989;15:233-51.
16. Isaacs JT. Antagonistic effect of androgen on prostatic cell death. Prostate 1984;5: 545-57.
17. Stanisic T, Sadlowski R, Lee C, Grayhack JT. Partial inhibition of castration-induced ventral prostate regression with actinomycin D and cycloheximide. Invest Urol 1978; 16:19-22.
18. Kyprianou N, Isaacs JT. Activation of programmed cell death in the rat ventral prostate after castration. Endocrinology 1988;122:552-62.
19. Kyprianou N, English HF, Isaacs JT. Activation of a Ca^{2+}-Mg^{2+}-dependent endonuclease as an early event in castration-induced prostatic cell death. Prostate 1988; 13:103-18.

20. Kyprianou N, Isaacs JT. Expression of transforming growth factor-β in the rat ventral prostate during castration-induced programmed cell death. Mol Endocrinol 1989; 3:1515-22.
21. Buttyan R, Zaker Z, Lochshin R, Wolgemuth D. Cascade induction of c-*fos*, c-*myc* and heat shock 70K transcripts during regression of rat ventral prostate gland. Mol Endocrinol 1988;2:650-7.
22. Montpetit ML, Lawless KR, Tenniswood M. Androgen repressed messages in the rat ventral prostate. Prostate 1986;8:25-36.
23. Conor J, Sawdzuk IS, Benson MC, et al. Calcium channel antagonists delay regression of androgen-dependent tissues and suppress gene activity associated with cell death. Prostate 1988;13:119-30.
24. Buttyan R, Olsson CA, Pintar J, et al. Induction of the TRPM-2 gene in cells undergoing programmed death. Mol Cell Biol 1989;9:3473-81.
25. Carter MF, Chung LWK, Coffey DS. The temporal requirements for androgens during the cell cycle of the prostate gland in urological research. King LR, Murphy GP, eds. New York: Plenum Press, 1972:27-38.
26. Coffey DS, Shimazaki J, Williams-Ashman HG. Polymerization of deoxyribonucleotides in relation to androgen-induced prostatic growth. Arch Biochem Biophys 1968;124:184-98.
27. Kyprianou N, Isaacs JT. Quantal relationship between prostatic dihydrotestosterone and prostatic cell content: Critical threshold concept. Prostate 1987;11:41-50.
28. Saltzman AG, Hiipakka RA, Chang C, Liao S. Androgen repression of the production of a 29 kilodalton protein and its mRNA in the rat ventral prostate. J Biol Chem 1987; 262:432-7.
29. Lee C, Sensibar JA. Protein of the rat prostate: Synthesis of new proteins in the ventral lobe during castration-induced regression. J Urol 1985;138:903-8.
30. Wyllie AH. Glucocorticoid induces in thymocytes a nuclease-like activity associated with the chromatin condensation of apoptosis. Nature 1980;284:555-6.
31. Umansky SR, Korol BA, Nelipovich PA. *In vivo* DNA degradation in thymocytes of γ-irradiated or hydrocortisone-treated rats. Biochem Biophys Acta 1981;655:9-17.
32. Cohen JJ, Duke RC. Glucocorticoid activation of a calcium-dependent endonuclease in thymocyte nuclei leads to cell death. J Immunol 1984;132:38-42.
33. Wyllie AH, Morris RG, Smith AL, Dunlop D. Chromatin cleavage in apoptosis: Association with condensed chromatin morphology and dependence on macromolecular synthesis. J Pathol 1984;142:67-77.
34. Kerr JFR, Wyllie AH, Currie AR. Apoptosis: A basic biological phenomenon with wide ranging implications in tissue kinetics. Br J Cancer 1972;26:239-57.
35. Kerr JFR, Searle J. Deletion of cells by apoptosis during castration-induced involution of the rat prostate. Virchowa Archiv B 1973;13:87-102.
36. Bruchovsky N, Wilson JD. The conversion of testosterone to 5 alpha-androstan-17 beta-ol-3-one by rat prostate *in vivo* and *in vitro*. J Biol Chem 1968;243:3012-21.
37. Elson SD, Browne CA, Thorburn GD. Identification of epidermal growth factor-like activity in human male reproductive tissues and fluids. J Clin Endocrinol Metab 1984;58:589-94.
38. Shikata H, Utsumi N, Hiramatsu M, Minami N, Nemoto N, Shikata. Immunohistochemical localization of nerve growth factor and epidermal growth factor in guinea pig prostate gland. Histochemistry 1984;80:411-3.

39. Jacobs SC, Story MT, Sasse J, Lawson RK. Characterization of growth factors derived from the rat ventral prostate. J Urol 1988;139:1106-10.
40. Nishi N, Matuo Y, Wade F. Partial purification of a major type of rat prostatic growth factor: Characterization of an epidermal growth factor-related mitogen. Prostate 1988;12:209-20.
41. Jacobs SC, Story MT. Exocrine secretion of epidermal growth factor by the rat prostate: effect of adrenergic agents, cholinergic agents, and vasoactive intestinal peptide. Prostate 1988;13:79-87.
42. Fowler JE, James LT, Lau LG, Mills SE, Mounzer A. Epidermal growth factor and prostatic carcinoma: An immunohistochemical study. J Urol 1988;139:857-61.
43. Gregory H, Willshire IR, Kavanagh JP, Blacklock NJ, Chowdury S. Urogastrone-epidermal growth factor concentrations in prostatic fluid of normal individuals and patients with benign prostatic hypertrophy. Clin Sci 1986;70:359-63.
44. Connolly JM, Rose DP. Secretion of epidermal growth factor and related polypeptides by the DU 145 human prostate cancer cell line. Prostate 1989;15:177-86.
45. Wilding G, Valverius E, Knabbe C, Gelmann EP. Role of transforming growth factor-α in human prostate cancer cell growth. Prostate 1989;15:1-12.
46. Traish AM, Wotiz HH. Prostatic epidermal growth factor receptors and their regulation by androgens. Endocrinology 1987;121:1461-7.
47. Fekete M, Redding TW, Comaru-Schally AM, et al. Receptors for luteinizing hormone-releasing hormone, somatostatin, prolactin, and epidermal growth factor in rat and human prostate cancers and in benign prostate hyperplasia. Prostate 1989;14:191-208.
48. Davies P, Eaton CL. Binding of epidermal growth factor by human normal, hypertropic, and carcinomatous prostate. Prostate 1989;14:123-32.
49. Maddy SQ, Chisholm GD, Haekins RA, Habib FK. Localization of epidermal growth factor receptors in the human prostate by biochemical and immunocytochemical methods. J Endocrinol 1987;113:147-53.
50. Fiorelli G, DeBellis A, Longo A, Natali A, Costantini A, Serio M. Epidermal growth factor receptors in human hyperplasic prostate tissue and their modulation by chronic treatment with a gonadotropin-releasing hormone analog. J Clin Endocrinol Metab 1989;68:740-3.
51. Schuurmans ALG, Bolt J, Mulder E. Androgens stimulate both growth rate and epidermal growth factor receptor activity of the human prostate tumor cell LNCaP. Prostate 1988;12:55-63.
52. Sitaras NM, Sariban E, Bravo M, Pantazis P, Antoniades HN. Constitutive production of platelet-derived growth factor-like proteins by human prostate carcinoma cell lines. Cancer Res 1988;48:1930-5.
53. Matuo Y, Nishi N, Tanaka H, Sasaki I, Isaacs JT, Wada F. Production of IGF-II-related peptide by an anaplastic cell line (AT-3) established from the Dunning prostatic carcinoma of rats. In Vitro Cell Dev Biol 1988;24:1053-6.
54. Matuo Y, Nishi N, Matsui S, Sandberg AA, Isaacs JT, Wada F. Heparin binding affinity of rat prostatic growth factor in normal and cancerous prostate: Partial purification and characterization of rat prostatic growth factor in the Dunning tumor. Cancer Res 1987;47:188-92.
55. Nishi N, Matuo Y, Kunitomi K, et al. Comparative analysis of growth factors in normal and pathologic human prostates. Prostate 1988;13:39-48.

56. Jacobs SC, Russell KL. Mitogenic factor in human prostate extracts. Urology 1980;16:488-91.
57. Story MT, Jacobs SC, Lawson RK. Partial purification of a prostatic growth factor. J Urol 1984;132:1212-5.
58. Nishi N, Matuo Y, Muguruma Y, Yoshitake Y, Nishikawa K, Wada F. A human prostatic growth factor (hPGF): Partial purification and characterization. Biochem Biophys Res Commun 1985;132:1103-9.
59. Story MT, Sasse J, Jacobs SC, Lawson RK. Prostatic growth factor purification and structural relationship to basic fibroblast growth factor. Biochemistry 1987;26: 3843-9.
60. Mydlo JH, Bulbul MA, Richon VM, Heston WDW, Fair WR. Heparin-binding growth factor isolated from human prostatic extracts. Prostate 1988;12:343-55.
61. Mansson P-E, Adams P, Kan M, McKeehan WL. Heparin-binding growth factor gene expression and receptor characteristics in normal rat prostate and two transplantable rat prostate tumors. Cancer Res 1989;49:2485-94.
62. Smith EP, Russell WE, French FS, Wilson EM. A form of basic fibroblast growth factor is secreted into the adluminal fluid of the rat coagulating gland. Prostate 1989; 12:353-65.
63. Gospodarowicz D, Neufeld G, Schweigerer L. Fibroblast growth factor. Mol Cell Endocrinol 1986;46:187-204.
64. Mydlo J, Michaeli J, Heston W, Fair W. Expression of basic fibroblast growth factor m-RNA in benign prostatic hyperplasia and prostatic carcinoma. Prostate 1988;13: 241-8.
65. Mori H, Maki M, Oishi K, et al. Increased expression of genes for basic fibroblast growth factor and transforming growth factor type β2 in human benign prostatic hyperplasia. Prostate 1990;16:71-80.
66. Matuo Y, Nishi N, Takasuka H, et al. Production and significance of TGF-β in AT-3 metastatic cell line established from the Dunning rat prostatic adenocarcinoma. Biochem and Biophys Res Commun 1990;2:840-7.
67. Kyprianou N, Isaacs JT. Identification of a cellular receptor for transforming growth factor-β in rat ventral prostate and its negative regulation by androgens. Endocrinology 1988;123:2124-31.
68. Cunha GR, Donjacour A. Stromal-epithelial interactions in normal and abnormal prostatic development. In: Current concepts and approaches to the study of prostate cancer. New York: Alan R. Liss, 1987:251-72.
69. McKeehan WL, Adams PS, Posser MP. Direct mitogenic effect of insulin, epidermal growth factor, glucocorticoid, cholera toxin, unknown pituitary factors and possibly prolactin, but not androgen, on normal rat prostate epithelial cells in serum-free primary cell culture. Cancer Res 1984;44:1998-2010.
70. Nishi N, Matuo Y, Nakamoto T, Wada F. Proliferation of epithelial cells derived from rat dorsolateral prostate in serum-free primary cell culture and their response to androgen. In Vitro Cell Dev Biol 1988;24:778-86.
71. Peehl DM, Stanley TA. Growth responses of normal, benign hyperplastic, and malignant human prostatic epithelial cells in vitro to cholera toxin, pituitary extract, and hydrocortisone. Prostate 1986;8:51-61.
72. Chaproniere DM, McKeehan WL. Serial culture of single adult human prostatic epithelial cells in serum-free medium containing low calcium and a new growth factor from bovine brain. Cancer Res 1986;46:819-24.

73. Merchant DJ, Clarke SM, Harris S. Primary explant culture: An in vitro model of the human prostate. Prostate 1983;4:523-42.
74. Stoningten OG, Hemmingsen H. Culture of cells as a monolayer derived from the epithelium of the human prostate. J Urol 1971;106:393-400.
75. Heatfield BM, Sanefugi H, Trump BF. Studies of carcinogenesis of human prostate, III. Long-term explant culture of normal prostate and benign prostatic hyperplasia: Transmission and scanning electron microscopy. J Natl Cancer Inst 1982;69:757-66.
76. Sanefugi H, Heatfield BM, Trump BF, Young JD Jr. Studies of carcinogenesis of human prostate, II. Long-term explant culture of normal prostate and benign prostatic hyperplasia: Light microscopy. J Natl Cancer Inst 1982;69:751-6.
77. Verhagen APM, Aadlers TW, Ramaekers FCS, Debruyne FMJ, Schalken JA. Differential expression of keratins in the basal and luminal compartments of rat prostatic epithelium during degeneration and regeneration. Prostate 1988;13:25-38.
78. Kabalin JN, Peehl DM, Stamey TA. Clonal growth of human prostatic epithelial cells is stimulated by fibroblasts. Prostate 1989;14:251-63.
79. Story MT, Livingston B, Baeten L, et al. Cultured human prostate-derived fibroblasts produce a factor that stimulates their growth with properties indistinguishable from basic fibroblast growth factor. Prostate 1989;15:355-65.
80. McKeehan WL, Adams PS, Fast D. Different hormonal requirement for androgen-independent growth of normal and tumor epithelial cells from rat prostate. In Vitro Cell Dev Biol 1987;23:147-52.
81. McKeehan WL, Adams PS. Heparin-binding growth factor/prostatropin attenuates inhibition of rat prostate tumor epithelial cell growth by transforming growth factor type beta. In Vitro Cell Dev Biol 1987;24:243-6.
82. Katz AE, Benson MC, Wise GJ, et al. Gene activity during the early phase of androgen-stimulated rat prostate regrowth. Cancer Res 1989;49:5889-94.
83. Coffey RJ, Leof EB, Shipley GD, Moses HL. Suramin inhibition of growth factor receptor binding and mitogenicity in AKR-2B cells. J Cell Physiol 1987;132:143-8.
84. Betsholtz C, Johnson A, Heldin CH, Westermark B. Efficient reversion of simian sarcoma virus—transformation and inhibition of growth factor-induced mitogenesis by suramin. Proc Natl Acad Sci USA 1986;83:6440-4.
85. Sato Y, Rifkin DB. Autocrine activities of basic fibroblast growth factor: Regulation of endothelial cell movement, plasminogen activator synthesis, and DNA synthesis. J Cell Biol 1988;107:1199-1205.
86. Neufeld G, Gospodarowicz D. Protamine sulfate inhibits mitogenic activities of the extracellular matrix and fibroblast growth factor, but potentiates that of epidermal growth factor. J Cell Physiol 1987;132:287-94.
87. Huang JS, Nishimura J, Huang SS, Deuel TF. Protamine inhibits platelet derived growth factor receptor activity but not epidermal growth factor activity. J Cell Biochem 1984;26:205-20.
88. Myers C. Three centers start phase 2 trials with suramin; NCI responses hold up. Clin Cancer Lett 1989;12:5-6.
89. Taylor S, Folkman J. Protamine is an inhibitor of angiogenesis. Nature 1982;297: 307-12.
90. Nagle RB, Ahmann FR, McDoniel KM, Paquin ML, Clark VA, Celniker A. Cytokeratin characterization of human prostatic carcinoma and its derived cell lines. Cancer Res 1987;47:281-6.
91. Ramaekers FCS, Verhagen APM, Isaacs JT, et al. Intermediate filament expression

and the progression of prostatic cancer as studied in the Dunning R-3327 rat prostatic carcinoma system. Prostate 1989;14:323-39.

92. Scott WW, Menon M, Walsh PC. Hormonal therapy of prostatic cancer. Cancer 1980;45:1929-36.
93. Isaacs JT, Hukka B. Nonrandom-involvement of chromosome 4 in the progression of rat prostatic cancer. Prostate 1988;13:165-88.
94. Isaacs JT, Isaacs WB, Feitz WFJ, Scheres J. Establishment and characterization of seven Dunning rat prostatic cancer cell lines and their use in developing methods for predicting metastatic abilities of prostatic cancers. Prostate 1986;9:261-81.
95. Jacobs SC. Spread of prostatic cancer to bone. Urology 1983:337-44.
96. Chackal-Roy M, Niemeyer C, Moore M, Zetter BR. Stimulation of human prostatic carcinoma cell growth by factors present in human bone marrow. J Clin Invest 1989:43-50.

13

Estrogen Regulation of Uterine Epidermal Growth Factor Receptor and Nuclear Proto-Oncogenes

D. S. Loose-Mitchell, R. M. Gardner, S. M. Hyder, J. L. Kirkland,* T. -H. Lin,* R. B. Lingham, V. R. Mukku, C. A. Orengo, U. R. Tipnis, and G. M. Stancel

Department of Pharmacology, University of Texas Medical School, Houston, and *Division of Endocrinology, Department of Pediatrics, Baylor College of Medicine, Houston, Texas

The effects of the ovarian steroids on growth of the reproductive tract have been recognized for many years. Receptors for estrogen and progesterone are present in such tissues as the uterus, and it seems clear that these sex steroids stimulate a variety of responses, including growth.

More recently, we and others have determined that the uterus contains several polypeptide growth factors and/or their cognate receptors. For example, the normal uterus produces epidermal growth factor (EGF) (1) and insulin-like growth factor-I (IGF-I) (2), and also contains receptors for EGF (3), platelet-derived growth factor (PDGF) (4), and IGF-I (5). Such observations raise the possibility that peptide growth factors may regulate uterine growth, independently and/or in concert with estrogens and progesterone.

Our laboratory has been interested in the control of uterine EGF receptor levels by estrogen. This interest stems in part from a number of observations that suggest a role for EGF in estrogen-mediated growth: (*a*) EGF is present in the uterus and in luminal fluid (1, 6–7); (*b*) EGF receptors are present in the uterus (3, 8–9); (*c*) EGF supports the growth of cultured cells derived from the uterus (10–11); and (*d*) perhaps most importantly, antibodies to EGF block estrogen-stimulated growth of uterine organ cultures (12).

Given a possible role of EGF in estrogen-stimulated uterine growth, we have characterized EGF receptors in this tissue and studied their regulation by estrogen. These studies are reviewed here and led us in turn to consider two other questions. The first was whether EGF was capable of stimulating myometrial contractility. The second was whether estrogen-stimulated growth involves the expression of nuclear proto-oncogenes. Our available data suggests that the answer to both questions is affirmative, and these studies are also reviewed here.

UTERINE EGF RECEPTORS

Properties of Receptors

EGF produces its effects by interacting with specific, high-affinity receptors present in target cells (13). As seen in Figure 1, membranes prepared from the immature rat uterus contain saturable, high-affinity EGF-binding sites. In a number of studies using different species (mouse and rat), different age animals (immature and mature), and different endocrine states, we have observed that uterine membranes contain only a single class of EGF-binding sites, with a Kd value in the range of 0.4- to 2-nM EGF. Membranes prepared from animals not exposed to estrogen generally contain about 200 fmoles of EGF-binding sites per milligram of membrane protein. Estrogen treatment in vivo increases the number of functional EGF receptors (see below), but does not alter the affinity of the receptor for the growth factor (8). These binding sites are specific for EGF since the binding of

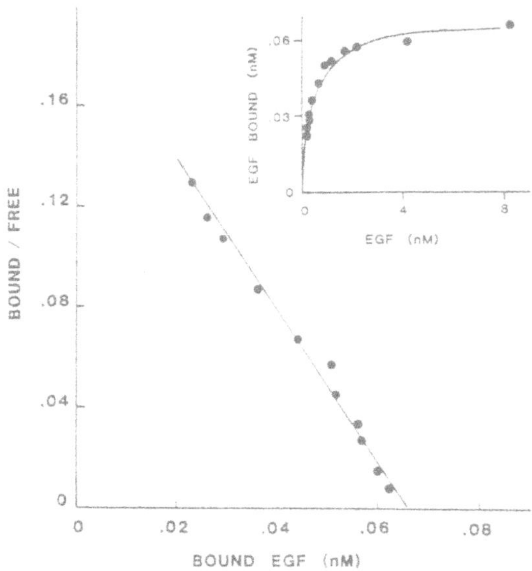

Fig. 1. The specific binding of ^{125}I-EGF to isolated uterine membranes. The binding is saturable (*see inset*) and Scatchard analysis (14) indicates a single class of binding sites with a Kd of 0.4 nM. (From Mukku VR, Stancel GM, Receptors for epidermal growth factor in the rat uterus, Endocrinology 1985;117:149–54, with permission.)

^{125}I-EGF is displaced by an excess of the unlabeled growth factor, but not by other peptides (3).

In the immature rat uterus, EGF-binding sites are observed in luminal and glandular epithelium, stroma, and myometrium using labeled ligand binding to tissue segments followed by autoradiography. A number of control studies revealed that these binding sites represent authentic EGF receptors (14). It is important to note that in the immature animal all these cell types undergo a growth response following estrogen administration in vivo (15). Similarly, Chegini, et al. (16) have used autoradiography to determine that all major cell types of the human uterus contain EGF receptors, and several groups have used ligand binding to illustrate that endometrial (9–10) and myometrial (9, 11) cells contain EGF receptors.

The mechanism of signal transduction of the EGF receptor appears to be activation of a tyrosine protein kinase activity upon ligand binding (13). We have shown that incubation of solubilized uterine membrane preparations with EGF stimulates autophosphorylation of the 170,000 MW EGF receptor (3, 8). Additional studies have shown that this phosphorylation occurs primarily on tyrosine residues of the receptor (3) and that EGF binding also stimulates kinase activity measured with an exogenous substrate containing a tyrosine residue (8). Chemical cross-linking studies with the bifunctional reagent dissuccinimidylsuberate have established also that the molecular weight of the uterine EGF receptor is 170,000 (3, 8). Taken together, these results illustrate that the uterine EGF receptor is similar to that reported in a wide variety of other tissues and cells (13).

Regulation of EGF Receptors by Estrogen

Administration of estradiol to immature animals produces a 2- to 3-fold increase in functional EGF receptors assessed by ligand binding (Fig. 2). This effect occurs primarily between 6 and 12 h after hormone treatment and is specific for estrogenic steroids (8). This increase occurs well before tissue DNA synthesis that begins to increase roughly 15 h after steroid administration and is not maximum until 21–24 h after hormone treatment (15). Estradiol treatment also produces a comparable increase in tyrosine kinase activity of the EGF receptor (8).

The studies illustrated in Figure 2 were performed with immature rats. More recent studies have established that estrogen increases uterine EGF receptor levels in the immature mouse (Gardner, et al., in preparation) and in the castrate adult rat (17). In addition, the level of uterine EGF receptor varies throughout the estrous cycle in rats in parallel with changes in plasma estrogens and occupied-tissue estrogen receptors (17). The generality of this effect suggests that the regulation of uterine EGF receptors by estrogen is a physiological effect.

The induction of uterine EGF receptors by estrogen is sensitive to both cycloheximide and actinomycin D (8), suggesting that the observed increases represent de novo receptor synthesis and that the mechanism of induction is at least partially transcriptional in nature. This possibility received further support from the demonstration that estrogen treatment in vivo increases the level of EGF receptor mRNA (18). As illustrated in Figure 3, estradiol treatment of immature

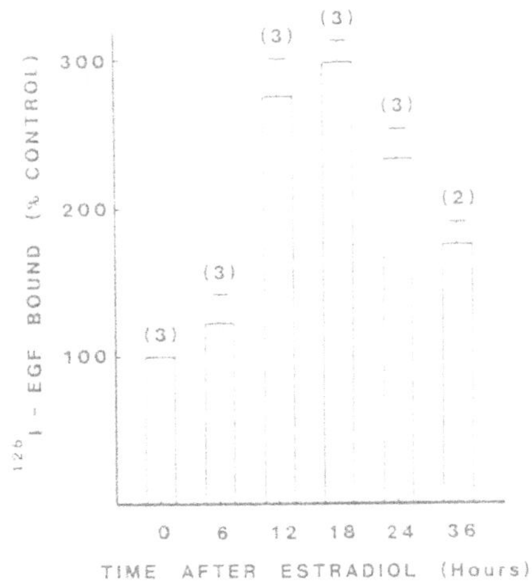

Fig. 2. Regulation of uterine EGF receptors by estrogen. Immature rats were treated with estradiol for the indicated times prior to sacrifice. Uterine membranes were prepared and analyzed for specific ^{125}I-EGF binding. (From Mukku VR, Stancel GM, Regulation of epidermal growth factor receptor by estrogen, J Biol Chem 1985;260:9820–4, with permission.)

rats leads to an increase in the 9.5-kb EGF receptor transcript prior to an increase in functional receptor levels. This increase in the EGF receptor mRNA is sensitive to actinomycin D but not puromycin, and the increase is specific for estrogenic steroids (18). We have also observed a similar increase in the mouse uterine EGF receptor mRNA following estradiol treatment in vivo (Gardner, et al., unpublished observation).

One interpretation of these results is that the induction of the EGF receptor is a primary effect of estradiol acting via its nuclear receptor. This mechanism is compatible with our existing data, but other experiments, such as direct measurements of transcription rates, evaluation of message stability, and a search for estrogen responsive elements in the EGF receptor gene region, will be required to unequivocally establish this point.

It is also conceivable that estrogen might act initially through a nontranscriptional mechanism to trigger the release of EGF from precursor sites (1), and the EGF thus formed might be the stimulus to induce the production of its own receptor (19–21). While we believe this possibility is less likely, it cannot be ruled out on the basis of the available data.

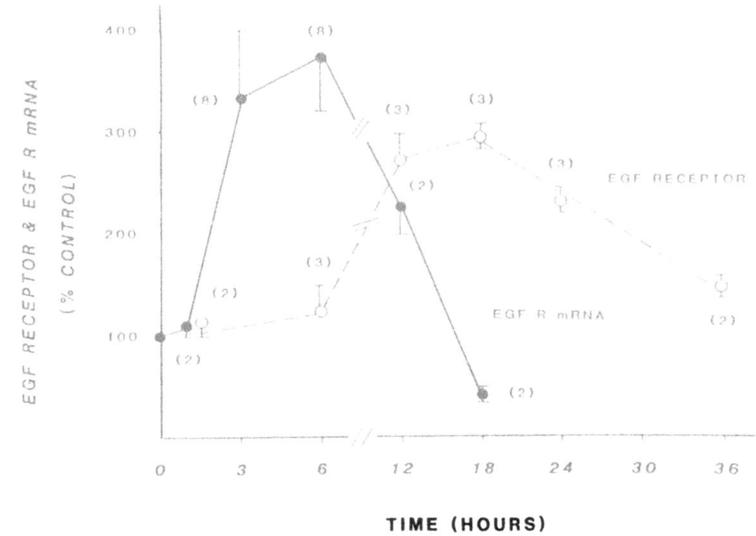

Fig. 3. Regulation of EGF receptor mRNA by estrogen. Immature rats were treated with estradiol for the indicated times prior to sacrifice. mRNA levels were determined by densitometric analysis of RNA blots hybridized with a riboprobe complementary to the rat EGF receptor cDNA. EGF receptor levels were determined by ligand binding. (From Lingham RB, Stancel GM, Loose-Mitchell DS, Estrogen regulation of epidermal growth factor receptor messenger RNA, Mol Endocrinol 1988;2:230–5, with permission.)

Relationship Between EGF Receptor Induction and Uterine DNA Synthesis

We are currently investigating the relationship between the increases in uterine EGF receptor levels and DNA synthesis following estrogen treatment. While these studies are not yet complete, available results include the following: (*a*) dose-response and hormonal specificity profiles for the two parameters are virtually identical; and (*b*) a single injection of short-acting estrogens does not appreciably increase receptor levels or DNA synthesis, but repeated administration of these compounds increases both to the same degree as estradiol treatment. At present, it appears that increases in tissue DNA synthesis after estrogen treatment correlate with, and are preceded by, increases in EGF receptor levels, although the evidence available does not necessarily indicate a direct cause-effect relationship.

One can envision two general ways that an elevation in growth factor receptor synthesis might play an important role in tissue DNA synthesis. Increases in the EGF receptor level, alone or in combination with increases in EGF or EGF-like ligands, might be necessary to produce a threshold level of a cellular second messenger required for progression towards, or initiation of, tissue DNA synthesis. Alternatively, cells stimulated to grow might require an increased synthesis of receptors to prevent down-regulation below the point necessary to sustain a prolonged signal required for DNA synthesis. Further studies are clearly required to determine the role of EGF receptor levels in estrogen-stimulated growth.

STIMULATION OF MYOMETRIAL CONTRACTIONS BY EGF

Most studies of EGF have focused on the growth-promoting effects of this peptide. However, it is clear that this peptide also has several other biological activities, including the ability to decrease gastric acid secretion (22–23) and to produce contractions of vascular smooth muscle (24–25). These results, coupled with the finding that the uterine myometrium contains EGF receptors (14, 16), led us to ask whether EGF might stimulate myometrial contractions.

The addition of EGF to segments of uterine tissue in an in vitro organ bath system rapidly stimulates contractile activity (Fig. 4). This effect is produced by low

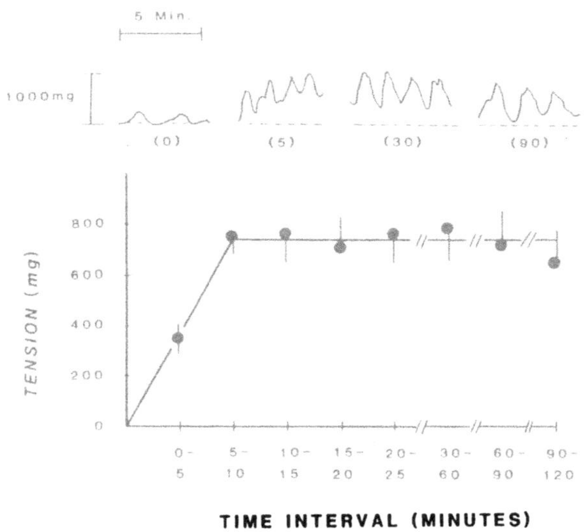

Fig. 4. Stimulation of uterine contractions by EGF. *Top*: EGF was added in vitro to an isolated uterine segment from a mature castrate animal that received estradiol priming in vivo for 24 h prior to sacrifice. The tracings represent contractile activity at the indicated times after EGF addition. *Bottom*: Composite showing the maximum tension developed in vitro in the indicated time intervals after EGF addition to a series of isolated uterine tissues; N = 6–7 per point. (From Gardner RM, Lingham RB, Stancel GM, Contractions of the isolated uterus stimulated by epidermal growth factor, FASEB J 1987;1:224–8, with permission.)

concentrations of EGF (ED_{50} of 3.5 nM) and is specific since it is not produced by other peptides, such as insulin or MSA (26). This response to EGF requires in vivo estrogen priming of the tissue, but it seems unlikely that this requirement is due solely to the induction of growth factor receptor by the steroid (26). Other studies have revealed that EGF stimulates the myometrium directly since contractions occur if the endometrium is physically removed from the muscle layer (27).

Recent results suggest that EGF causes the release of arachidonic acid from uterine membrane sites. The arachidonic acid then appears to be converted to both prostaglandins and leukotrienes, which are well-known stimulators of myometrial contractility (27). This possible mechanism for the production of uterine contractions by EGF is tentative, however, since it is based on the use of pharmacological inhibitors rather than on direct measurements of arachidonic acid metabolites (27). It is clear, nevertheless, that EGF is a potent and efficacious stimulant of myometrial contractions, and this raises the clear possibility that EGF may produce physiological effects on the myometrium other than growth.

REGULATION OF NUCLEAR PROTO-ONCOGENES BY ESTROGEN

It seemed logical at this point to next investigate uterine responses that could conceivably be activated via an EGF receptor-dependent pathway. At about this time, virologists had made a number of breakthroughs in identifying retroviral oncogenes, and it soon became obvious that (*a*) Nontransformed cells contained analogs (so-called proto-oncogenes or cellular oncogenes) of viral oncogenes; (*b*) some of these proto-oncogenes were either growth factors or growth factor receptors; and (*c*) some growth factors stimulated proto-oncogene expression (28–29). These findings suggested that proto-oncogenes and growth factors play an important role in the growth and function of normal cells. In particular, several groups demonstrated that EGF stimulated the expression of the proto-oncogene c-*fos* in a variety of cultured cells (30).

For these experiments animals were treated with estrogen, and total uterine RNA samples were prepared after various times. As seen in Figure 5, there is a very large, rapid increase in c-*fos* expression measured by blot analysis. Much to our surprise, this increase clearly preceded the increase in functional EGF receptor levels previously observed (see Fig. 3). In fact, the rapidity of this induction clearly suggested that estrogen was directly regulating c-*fos* expression. In conjunction with these studies, in situ hybridization experiments have shown that all major uterine cell types (luminal and glandular epithelium, stroma, and myometrium) contain c-*fos* transcripts. The level of the messenger RNA for this proto-oncogene is also increased in all of these cell types following estrogen treatment in vivo (Tipnis, et al., in preparation). This is consistent with a role for *fos* in cell growth since all uterine cell types initiate DNA synthesis following estrogen administration to the immature animal (15).

A direct effect of estrogen on c-*fos* expression was suggested by the following observations: (*a*) Induction is blocked by actinomycin D, but not puromycin (31);

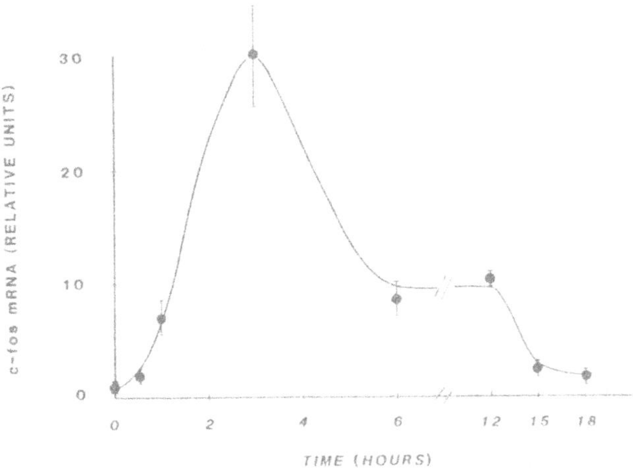

Fig. 5. Regulation of c-*fos* mRNA by estrogen. Immature rats were treated with estradiol for the indicated times prior to sacrifice. Total uterine RNA was prepared and analyzed by blot analysis. Values indicate relative levels of c-*fos* mRNA determined by densitometry. (From Loose-Mitchell DS, Chiappetta C, Stancel GM, Estrogen regulation c-*fos* messenger ribonucleic acid, Mol Endocrinol 1988;2:946–51, with permission.)

(*b*) estrogen leads to an increase in c-*fos* expression measured by nuclear run on measurements (32); and (*c*) the 5'-upstream region of the c-*fos* gene contains a palindromic sequence with a high degree of homology to the consensus sequence for the estrogen-responsive element, or ERE (31). In recent transfection experiments, we have been able to demonstrate that this 5'-upstream region of c-*fos* confers estrogen inducibility to a CAT-reporter construct (33).

OVERALL UTERINE GROWTH IN RESPONSE TO ESTROGEN

It has been recognized for some time that the regulation of uterine growth by estrogens is a process that is regulated in several stages and requires more than an initial, transient interaction of the estrogen receptor with an agonist. Endocrinologists have referred to these stages of the estrogen-induced growth response as early effects and late effects (34–35). More recently, we have been struck by the general similarity between this model of uterine growth and models of growth in cultured fibroblasts.

In the fibroblast model the mitogenic stimulation of quiescent cells (in G_0) occurs in at least two major steps. The first has been termed *competence* and is thought to represent the movement of arrested cells from G_0 into the early portion of G_1, or at least to the G_0/G_1 interface. Movement of cells through G_1 toward S is termed *progression*. In the fibroblast system there is a clear demarcation between competence and progression because the two stages are controlled by different peptide factors. PDGF is the prototype competence factor, but alone cannot stimulate mitogenesis. Cells must first be exposed to PDGF and then to a progression factor, such as EGF or IGF-I (28, 36). In the uterine system, however, this distinction is not as sharp because estrogen itself seems to be involved in both early and late effects (34–35).

The analogy between the two systems is even more striking if one considers that PDGF regulates *fos* and *myc* expression in fibroblasts (37) and that estradiol regulates *fos* (31–32) and *myc* (38–39) expression in the uterus. While the two stimuli have in common the regulation of these specific genes, mechanistic steps between receptor-agonist interaction and gene expression are clearly different for PDGF and the steroid. Similarly, EGF and IGF-I are progression factors for fibroblasts, and estradiol seems to increase the expression of IGF-I (2), the level of EGF (1), the level of EGF receptors (8, 18), and the level of IGF-I receptors (5) in the uterus. It thus seems reasonable to suggest that the so-called early and late phases of estrogen action may be analogous in a general way to competence and progression in fibroblasts.

If one considers *fos* expression as a marker of the competence state, estrogenic steroids may function via their receptors to elicit competence responses directly at the genomic level. Given our data and the work of others, this seems the most reasonable possibility. While it seems less likely to us, estrogens could conceivably act via a nongenomic mechanism (e.g., conversion of an inactive EGF precursor to an active peptide) that functions through a membrane receptor to stimulate c-*fos* expression. PDGF-like regulation, on the other hand, involves a cytoplasmic second messenger(s) to transmit a signal emanating from the plasma membrane. The basic idea is that steroids and polypeptide growth factors control competence by different regulatory mechanisms that converge at a common end point(s), such as expression of c-*fos*.

Possible mechanistic analogies between progression-like events in the two systems are more difficult to envision. An obvious possibility is that the estrogen-receptor complex directly controls the local production of such factors as EGF/IGF-I, and/or their receptors, and these act via autocrine or paracrine mechanisms to control transit through G_1. Unfortunately, it is difficult to evaluate this possibility given existing data on the mechanism of regulation of EGF and IGF-I by estrogens. Additional studies will obviously be required to evaluate this possibility and to understand the in vivo role of proto-oncogenes and growth factors in estrogen-controlled uterine growth.

REFERENCES

1. DiAugustine RP, Petrusz P, Bell GI, et al. Influence of estrogens on mouse uterine epidermal growth factor precursor protein and messenger RNA. Endocrinology 1988; 122:2355-63.

2. Murphy CJ, Murphy CC, Friesen HG. Estrogen induces insulin-like growth factor-1 expression in the uterus. Mol Endocrinol 1987;1:445-50.
3. Mukku VR, Stancel GM. Receptors for epidermal growth factor in the rat uterus. Endocrinology 1985;117:149-54.
4. Ronnstrand L, Beckmann MP, Faulders B, Ostman A, Ek B, Heldin CH. Purification of the receptor for PDGF from porcine uterus. J Biol Chem 1987;262:2929-32.
5. Ghahary A, Murphy CJ. Uterine insulin-like growth factor-1 receptors: regulation by estrogen and variation throughout the estrous cycle. Endocrinology 1989;125: 597-604.
6. Gonzalez F, Lakshmanan J, Hoath S, Fisher DA. Effect of oestradiol-17B on uterine epidermal growth factor concentration in immature mice. Acta Endocrinol 1984; 105:425-8.
7. Imai Y. Epidermal growth factor in rat uterine luminal fluid [Abstract]. Endocrinology 1982;110(suppl):162.
8. Mukku VR, Stancel GM. Regulation of epidermal growth factor receptor by estrogen. J Biol Chem 1985;260:9820-4.
9. Hofmann GE, Rao CV, Barrows GH, Sanfilippo JS. Binding sites for epidermal growth factor in human uterine tissues and leiomyomas. J Clin Endocrinol Metab 1984;58:880-4.
10. Tomooka Y, DiAugustine RP, McLachlan JA. Proliferation of mouse uterine epithelial cells in vitro. Endocrinology 1986;118:1011-8.
11. Bhargava G, Rifas L, Makman MH. Presence of epidermal growth factor receptors and influence of epidermal growth factor on proliferation and aging in cultured smooth muscle cells. J Cell Physiol 1979;100:365-74.
12. McLachlan JA, DiAugustine RP, Newbold RR. Estrogen induced uterine cell proliferation in organ culture is inhibited by antibodies to epidermal growth factor [Abstract]. Prog of the 69th meet of the Endocr Soc, Indianapolis 1987;313:99.
13. Carpenter G. Receptors for epidermal growth factor and other polypeptide mitogens. Annu Rev Biochem 1987;56:881-914.
14. Lin TH, Mukku VR, Verner G, Kirkland JL, Stancel GM. Autoradiographic localization of epidermal growth factor receptors to all major uterine cell types. Biol Reprod 1988;38:403-11.
15. Kaye AM, Sheratzky D, Lindner HR. Kinetics of DNA synthesis in immature rat uterus: age dependence and estradiol stimulation. Biochem Biophys Acta 1972; 261:475-86.
16. Chegini N, Rao CV, Barrows GH, Sanfilippo JS. Binding of ^{125}I-epidermal growth factor in human uterus. Cell Tissue Res 1986;246:543-8.
17. Gardner RM, Verner G, Kirkland JL, Stancel GM. Regulation of uterine epidermal growth factor (EGF) receptors by estrogen in the mature rat and during the estrous cycle. J Steroid Biochem 1989;32:339-43.
18. Lingham RB, Stancel GM, Loose-Mitchell DS. Estrogen regulation of epidermal growth factor receptor messenger RNA. Mol Endocrinol 1988;2:230-5.
19. Earp HS, Austin KS, Blaisdell J, et al. Epidermal growth factor (EGF) stimulates EGF receptor synthesis. J Biol Chem 1986;261:4777-80.
20. Clark AJL, Ishii S, Richert N, Merlino GT, Pastan I. Epidermal growth factor regulates the expression of its own receptor. Proc Natl Acad Sci USA 1985;82:8374-8.
21. Kudlow JE, Cheung CYM, Bjorge JD. Epidermal growth factor stimulates the syn-

thesis of its own receptor in human breast cancer cell line. J Biol Chem 1986;261: 4134-8.

22. Carpenter G, Cohen S. Epidermal growth factor. Annu Rev Biochem 1979;48:193-216.
23. Gregory H. In vivo aspects of urogastrone-epidermal growth factor. J Cell Sci 1985; 3(suppl):11-7.
24. Muramatsu I, Hollenberg MD, Lederis K. Vascular actions of epidermal growth factor-urogastrone: Possible relationship to prostaglandin production. Can J Physiol Pharmacol 1985;63:994-9.
25. Berk BC, Brock TA, Webb RC, et al. Epidermal growth factor, a vascular smooth muscle mitogen, induces rat aortic contraction. J Clin Invest 1985;75:1083-6.
26. Gardner RM, Lingham RB, Stancel GM. Contractions of the isolated uterus stimulated by epidermal growth factor. FASEB J 1987;1:224-8.
27. Gardner RM, Stancel GM. Blockade of epidermal growth factor induced uterine contractions by indomethacin or nordihydroguaritic acid. J Pharmacol Exp Therapeutics 1989;250:882-6.
28. Deuel TF. Polypeptide growth factors: Roles in normal and abnormal cell growth. Annu Rev Cell Biol 1987;3:443-93.
29. Bishop JM. Viral oncogenes. Cell 1985;42:23-38.
30. Alt FW, Harlow E, Ziff EB. Nuclear oncogenes. Cold Spring Harbor, NY: Cold Spring Harbor Lab., 1987.
31. Loose-Mitchell DS, Chiappetta C, Stancel GM. Estrogen regulation c-*fos* messenger ribonucleic acid. Mol Endocrinol 1988;2:946-51.
32. Weisz A, Bresciani F. Estrogen induces expression of c-*fos* and c-*myc* protooncogenes in rat uterus. Mol Endocrinol 1988;2:816-24.
33. Hyder SM, Stancel GM, Loose-Mitchell DS. Mapping of an estrogen response element within the 5'-flanking region of the mouse c-*fos* gene. J Cell Biochem 1989; 14B(suppl):233.
34. Anderson JN, Clark JH, Peck EJ Jr. The relationship between nuclear receptor estrogen binding and uterotrophic responses. Biochem Biophys Res Commun 1972; 48:1460-8.
35. Anderson JN, Peck EJ Jr, Clark JH. Nuclear receptor estrogen complex: Relationship between concentration and early uterotrophic responses. Endocrinology 1973;92: 1488-95.
36. Stiles CD, Capone GT, Scher CD, Antoniades HN, Van Wyk JJ, Pledger WJ. Dual control of cell growth by somatomedins and platelet-derived growth factor. Proc Natl Acad Sci 1979;76:1279-83.
37. Rollins BJ, Stiles CD. Regulation of c-*myc* and c-*fos* protooncogene expression by animal cell growth factors, in vitro. Cell Dev Biol 1988;24:81-4.
38. Travers MT, Knowler JT. Oestrogen induced expression of oncogenes in the immature rat uterus. FEBS Lett 1987;211:27-30.
39. Murphy LJ, Murphy LC, Friesen HG. Estrogen induction of N-*myc* and c-*myc* protooncogene expression in the rat uterus. Endocrinology 1987;120:1882-8.

14

Estrogens and Growth Factors in the Development, Growth, and Function of the Female Reproductive Tract

J. A. McLachlan, K. G. Nelson, T. Takahashi, N. L. Bossert, R. R. Newbold, and K. S. Korach

Laboratory of Reproductive and Developmental Toxicology, National Institute of Environmental Health Sciences, National Institutes of Health, Research Triangle Park, North Carolina

The mammalian uterus undergoes significant growth on a regular basis. For example, in a mouse within an estrous cycle of four days, the number of uterine epithelial cells must double within two days and return to normal levels in the next two. This process is repeated throughout the reproductive lifetime of the animal, which means that the murine uterine epithelium will be reconstituted approximately 90 times in one year. This remarkable growth potential obviously requires numerous controls. A single prime stimulus for initiation of uterine epithelial proliferation is a female sex hormone, estrogen, in the appropriate pharmacological form. Although estrogen is the apparent proximate effector of uterine epithelial cell division, the actual mechanism whereby its mitogenic signal is transduced within the uterine tissue or cell is still not completely known. Certainly, at the molecular level, the estrogen receptor (ER), functioning as a transcription factor, is involved in regulation of expression of specific genes, some of which may be involved in estrogen-induced mitogenesis; these pathways, however, remain to be established. At the same time, results have been presented that raise the possibility that peptide growth factors (or polyfunctional regulating factors) are involved in uterine cell biology.

Acknowledgments: The preparation of this manuscript by Mrs. Ann Marie Steffen and Ms. Vickie Englebright is gratefully acknowledged. The patience of Drs. Lisa Kern and David Schomberg is also appreciated. I would like to acknowledge Dr. Charles Daniel for framing the criteria used to evaluate growth factor physiology.

UTERINE GROWTH AND GROWTH FACTORS

Numerous growth factors and their receptors have been reported in mammalian uterine tissue. Their functions in uterine cell biology are largely speculative at present; the same can be said for their relationship to estrogen action. The purpose of this short report is to examine selectively some of the evidence currently available and to describe a framework in which to evaluate it.

Epidermal growth factor (EGF) and its receptor have been reported in uterine tissue and is described in more detail below. Transforming growth factor α (TGFα) has been described in the decidual portion of the rat uterus, but not in the intervening undecidualized tissue (1); moreover, in the same study TGFα mRNA was not found either by Northern analysis or in situ hybridization in uterine tissue before implantation. Both of these peptide factors, EGF and TGFα, are reported mitogens for various cells in culture and apparently share the same receptor (2).

Insulin-like growth factor I (IGF-I) is a polypeptide that shares sequence homology with insulin; it is only weakly mitogenic for cells in culture, although it potentiates the mitogenic effects of EGF (3). IGF-I protein and mRNA have been demonstrated in the rat uterus; IGF-I levels are increased in the uterus following treatment with estradiol, while no change in expression was seen in liver or kidney (4–5). Moreover, receptors for IGF-I were found in the uterus and were also shown to be under estrogen control (6). A function for uterine IGF-I has not yet been described, but the presence of both ligand and receptor raises the possibility for biologic function. This possibility is further supported by the observation that an IGF-I-binding protein is a major secretory product of the decidualized human endometrium (7).

A uterine growth factor for which a function has been proposed is colony-stimulating factor I (CSF-I). The postimplantation murine uterus expresses CSF-I mRNA (8), and the pregnant mouse uterus at term has CSF-I protein concentrations 1000-fold higher than the nonpregnant uterus (9). The fact that, among other considerations, the CSF-I receptor mRNA was first detected in the uterine decidua and, subsequently, in the mature placental trophectoderm has led Pollard (10) to suggest that CSF-I plays a role in murine placentation and accumulation of uterine macrophages.

The transforming growth factor βs (TGFβs) are a family of polypeptides with both stimulatory and inhibitory actions on cell division in culture (11). $TGF\beta_1$, a member of this family, was localized immunocytochemically in the periglandular connective tissue of the mouse uterus (12). Recently, the localization and timing of $TGF\beta_1$ expression in the mouse uterus during the preimplantation stages of pregnancy were determined by immunocytochemistry and in situ and Northern blot hybridization analysis (13). These workers report that the $TGF\beta_1$ message was localized in uterine epithelial cells prior to implantation, while at least one form of the peptide was highly localized to the extracellular space around the stroma (especially during early decidualization). The authors conclude that in the early pregnant mouse uterus, TGFβ is a product of the epithelial cell that signals stromal differentiation. This attractive hypothesis awaits further experimental confirmation.

Thus, while the mammalian uterus contains many growth factors, their functional significance is not yet known. In order to establish potential biological importance for a growth factor, several critera that were alluded to above should be met: (*a*) The endogenous ligand should be present in the appropriate form and site; (*b*) a high-affinity receptor for that ligand should be localized in the appropriate cells; (*c*) administration of the exogenous ligand should produce the appropriate physiological response; and (*d*) neutralization of the ligand in situ should cause reversal of the predicted or expected physiological effect. Work from our laboratory and others on EGF in the uterus fulfills most of these criteria. These are briefly outlined below.

THE CASE FOR EGF INVOLVEMENT IN UTERINE CELL BIOLOGY

EGF Ligand Present in Uterus

EGF-like activity was reportedly elevated in the mouse uterus following estrogen treatment (14). To demonstrate the presence of a putative EGF ligand in the uterus, our laboratory reported the levels of preproEGF mRNA in the mouse uterus and described a small (2-fold) increase in this message following estrogen treatment (15); the message levels were always low compared to submaxillary gland or kidney. Recently, Huet-Hudson, et al. (16) described a greater induction of preproEGF mRNA in estrogen-treated mice and further localized the message by in situ hybridization to the uterine epithelium. These results taken together establish the presence of the mRNA for the EGF ligand in the uterus, probably in the epithelial cell.

Immunoactivity for EGF can also be demonstrated in the uterine epithelial cell for the proform in ovariectomized and estrogen-treated mice (15) or for the mature form only following estradiol (16). In uterine tissue fragments, radioactively labeled in vitro, EGF synthesis was shown (16). Finally, mature EGF ligand was identified by radioimmunoassay in uterine luminal fluid of estrogen-treated mice; the levels of immunorecognizable EGF in this secreted form in vivo was in the range of 1 to 3 ng/mL (15, 17). Thus, in addition to the message for preproEGF, the estrogen-stimulated uterine epithelial cell contains the pro or mature forms of EGF (or both). Further, in vivo estrogen treatment is followed by secretion or release of free ligand into the luminal fluid.

EGF Receptor in Uterus

A high-affinity receptor for EGF was demonstrated in the rat uterus (18); the levels of bound EGF increased with time following estrogen treatment (19). Both the protein and mRNA for EGF receptor appear to be up-regulated by estrogen in the rat uterus (20). EGF receptors have also been identified in the uterus of the mouse (21), and human (22–24). In the mouse (21), rat (25), and human (22–23), I^{125}-EGF binds to all the major uterine cell types (epithelium, stroma, and myometrium). Recent work from our laboratory has also demonstrated that EGF receptor is an ontogenically early constituent of the developing uterus, detectable in the mullerian duct as early as day 13 of pregnancy (21). On the contrary, the ER is not detectable in the mouse uterine epithelium until day 4 postnatally (26). Thus, the EGF receptor is detected much earlier in female genital tract development than the ER. Since these ER-deficient cells are still responsive to estrogen-associated mitogenic signaling (27), this raises the possibility that EGF or its receptor may be involved in transducing the estrogen stimulus.

Physiological Effects Produced by Exogenous EGF

Murine uterine epithelial cells cultured on collagen gels in serum-free media respond mitogenically to EGF preferentially to various other growth factors; the optimal EGF concentration for enhanced cell proliferation was 10 ng/mL (28). Likewise, EGF stimulated proliferation of rabbit endometrial cells when it was added to the culture media (29). These studies establish the uterine epithelial cell as a target for the mitogenic influence of EGF. Other studies in our laboratory have shown that EGF stimulated proliferation and differentiation of the murine uterus in organ culture in a way that was similar to estrogen-treated uteri (McLachlan and Newbold, in preparation).

A simple in vivo test for the physiological responses associated with exogenous EGF is seen in the recent experiments conducted by Dr. Karen G. Nelson in our group (17). In these studies pellets containing EGF were placed in the subrenal capsule of ovariectomized immature mice, and the proliferative response in various tissues in situ was determined by their tritiated thymidine labeling index (LI). Less than 5% of the uterine epithelial cells in ovariectomized mice incorporated tritiated thymidine (LI < 5%). Following treatment with estrogen, the LI increased 10-fold in the uterus. Strikingly, when mice were treated with EGF pellets alone, in the absence of estrogen, the uterine epithelial LI approached 80%. The responses of the uterus and vagina, both in terms of cell proliferation and differentiation, were remarkably similar to those seen following estrogen. These results taken together demonstrate that the uterine epithelial cell in vitro, in vivo, and in situ represents an important target for the mitogenic effect of EGF and that in vivo, EGF gives remarkably estrogen-like responses in the female genital tract.

Neutralization of EGF Response

Organ cultures of murine uterus respond to either estrogen or EGF with an approximately 2-fold increase in epithelial cell numbers; the proliferative response to estrogen in these cultures is blocked by the presence of antibodies to EGF in the media (McLachlan and Newbold, in preparation). Similarly, when EGF antibodies are introduced into mice by slow-release pellets, the mitogenic effect of estrogen on uterine epithelial cells in vivo was blocked approximately 70% (17).

Thus, the four principle criteria that establish the potential biologic significance of a growth factor for a tissue are largely met for EGF and the uterus. Furthermore, these experiments also raise the possibility that estrogen action, at least in the murine uterus, may involve growth factor mediation. Many questions remain to be answered: Does estrogen stimulate EGF through a processing step involving protease activation and a membrane-associated prohormone, or are EGF levels regulated transcriptionally? Are many different growth factors (at least

EGF, TGFα, IGF-I, and TGFβ) involved in estrogen action in the uterus, and if so, do they work sequentially or simultaneously? Certainly, just these two questions should provoke many interesting experiments—and there are obviously many more questions.

ESTROGEN ACTION, ESTROGEN RECEPTORS, GROWTH FACTORS, AND UTERINE ONTOGENY

Perhaps the greatest significance of the principles outlined above lies in understanding estrogen action in undifferentiated cells and tissues. It has been shown by our laboratory (26–27) and others (30–31) that the uterine epithelium of newborn mice is apparently deficient in ER. However, during this period, estrogen administration can stimulate epithelial cell proliferation (27, 30–31) and ER appearance (27). One explanation (30–31) suggests that the ER-rich stromal cells transduce the estrogen signal to the epithelium. Another possibility is that the level of receptor in the neonatal uterine epithelium is so low that it is undetectable by current techniques (26–27).

Still another possibility is raised by the appearance of EGF receptor and peptide long before the appearance of ER (21). That is, growth factors and ER may work concertedly to transduce the estrogen signal in the epithelium, especially at a time when the target tissues express heterogeneity with regard to ER-containing cells. Whether paracrine or autocrine loops are operational in estrogen target tissues is still an open question. However, principles established in the relatively undifferentiated newborn uterine epithelium may be pertinent to other putative stem cell populations in adult tissues.

REFERENCES

1. Han VKM, Hunter ES, Pratt RM, Zendegui JG, Lee DC. Expression of rat transforming growth factor α mRNA during development occurs predominantly in maternal decidua. Mol Cell Biol 1987;7:2335-43.
2. Massague J. Epidermal growth factor-like transforming growth factor II. Interaction with epidermal growth factor receptors in human placental membranes and A431 cells. J Biol Chem 1983;258:13614-20.
3. Corps AN, Brown KD. Ligand-receptor interactions involved in the stimulation of Swiss 3T3 fibroblasts by insulin-like growth factor. Biochem J 1988;252:119-25.
4. Murphy LJ, Murphy LC, Friesen HG. Estrogen induces insulin-like growth factor expression in the rat uterus. Mol Endocrinol 1987;1:445-50.
5. Murphy LJ, Friesen HG. Differential effects of estrogen and growth hormone on uterine and hepatic insulin-like growth factor-I gene expression in the ovariectomized hypophysectomized rat. Endocrinology 1988;122:325-32.
6. Ghahary A, Murphy LJ. Uterine insulin-like growth factor-I receptors: Regulation by estrogen and variation throughout the estrous cycle. Endocrinology 1989;125: 597-604.
7. Bell SC. Endometrial IGF-binding protein: A paracrine role in controlling endometrial or trophoblast growth. Res Reprod 1988;20:3.
8. Arceci RJ, Shanahan F, Stanley ER, Pollard JW. Temporal expression and location of

colony stimulating factor-1 (CSF-1) and its receptor in the female reproductive tract are consistent with CSF-1 regulated placental development. Proc Natl Acad Sci USA 1989;86:8818-22.

9. Bartocci A, Pollard JW, Stanley ER. Regulation of colony stimulating factor-1 during pregnancy. J Exp Med 1986;164:956-61.
10. Pollard JW. Regulation of polypeptide growth factor synthesis and growth factor-related gene expression in the rat and mouse uterus before and after implantation. J Reprod Fertil 1990;88:721-31.
11. Moses HL, Tucker RF, Leof EB, Coffey RJ, Halper J, Shipley GD. Type-beta transforming growth factor is growth stimulator and growth inhibitor. In: Feramisco J, Ozanne B, Stiles C, eds. Cancer cells; vol 3. Cold Spring Harbor, NY: Cold Spring Harbor Laboratory, 1985:65-71.
12. Thompson NL, Flanders KC, Smith JM, Ellingsworth LR, Roberts AB, Sporn MB. Expression of transforming growth factor-β1 in specific cells and tissues of adult and neonatal mice. J Cell Biol 1989;108:661-9.
13. Tamada H, McMaster MT, Flanders KC, Andrews GK, Dey SK. Cell type-specific expression of TGF-β1 in the mouse uterus during the periimplantation period. Mol Endocrinol 1990 (in press).
14. Gonzales F, Lakshmanan J, Hoath S, Fisher DA. Effect of estradiol-17β on uterine epidermal growth factor concentrations in immature mice. Acta Endocrinol Copenh 1984;105:425-8.
15. DiAugustine RP, Petrusz P, Bell GI, et al. Influence of estrogens on mouse uterine epidermal growth factor precursor protein and messenger ribonucleic acid. Endocrinology 1988;122:2355-63.
16. Huet-Hudson YM, Chakraborty C, De SK, Suzuki Y, Andrews GK, Dey SK. Estrogen regulates synthesis of EGF in mouse uterine epithelial cells. Mol Endocrinol 1990;4:510-23.
17. Nelson KG, Takahashi T, Bossert NL, et al. Growth factors replace estrogen in the stimulation of female genital tract growth and differentiation. Proc Natl Acad Sci USA 1990 (in press).
18. Mukku VR, Stancel GM. Receptors for epidermal growth factors in the rat uterus. Endocrinology 1985;117:149-54.
19. Mukku VR, Stancel GM. Regulation of epidermal growth factors by estrogens. J Biol Chem 1985;260:9820-4.
20. Lingham RB, Stancel GM, Loose-Mitchell DS. Estrogen regulation of epidermal growth factor receptor messenger ribonucleic acid. Mol Endocrinol 1988;2:230-5.
21. Bossert NL, Nelson KG, Ross KA, Takahashi T, McLachlan JA. Epidermal growth factor binding and receptor distribution in the mouse reproductive tract during development. Dev Biol 1990 (in press).
22. Hofmann GE, Rao CV, Barrows GH, Schultz GS, Sanfilippo JS. Binding sites for epidermal growth factor in human uterine tissues and leiomyomas. J Clin Endocrinol 1984;58:880-7.
23. Chegini N, Rao CV, Wakim N, Sanfilippo J. Binding of ^{125}I-epidermal growth factor in human uterus. Cell Tissue Res 1986;246:543-8.
24. Damjanov I, Mildner B, Knowles BB. Immunohistochemical localization of the epidermal growth factor receptor in normal human tissues. Lab Invest 1986;55:588-92.
25. Lin T-H, Mukku VR, Verner G, Kirkland JL, Stancel GM. Autoradiographic localiza-

tion of epidermal growth factor receptors to all major uterine cell types. Biol Reprod 1988;38:403-11.
26. Yamashita S, Newbold RR, McLachlan JA, Korach KS. Developmental pattern of estrogen receptor expression in female mouse genital tracts. Endocrinology 1989; 125:2888-96.
27. Yamashita S, Newbold RR, McLachlan JA, Korach KS. The role of the estrogen receptor in uterine epithelial proliferation and cytodifferentiation in neonatal mice. Endocrinology 1990 (in press).
28. Tomooka Y, DiAugustine RP, McLachlan JA. Proliferation of mouse uterine epithelial cells in vitro. Endocrinology 1986;118:1011-8.
29. Gerschenson LE, Conner EA, Yang J, Anderson M. Hormonal regulation of proliferation in two populations of rabbit endometrial cells in culture. Life Sci 1979; 24:1337-43.
30. Bigsby RM, Cunha GR. Estrogen stimulation of deoxyribonucleic acid synthesis in uterine epithelial cells which lack estrogen receptors. Endocrinology 1986;119:390-5.
31. Taguchi O, Bigsby RM, Cunha GR. Estrogen responsiveness and the estrogen receptor during development of the murine female reproductive tract. Develop Growth Differ 1988;30:301-3.

PART

V

EMBRYO-MATERNAL SIGNALING

15

Expression and Function of Growth Factor Ligands and Receptors in Preimplantation Mouse Embryos

Daniel A. Rappolee,[1] Karin S. Sturm,[1] Gilbert A. Schultz,[2] Claudio A. Basilico,[3] Daniel Bowen-Pope,[4] Roger A. Pedersen,[1] and Zena Werb[1]

[1]Laboratory of Radiobiology and Environmental Health and the Department of Anatomy, University of California, San Francisco; [2]Department of Medical Biochemistry, University of Calgary, Alberta, Canada; [3]Department of Pathology and Kaplan Cancer Center, New York University School of Medicine, New York; [4]Department of Pathology, University of Washington School of Medicine, Seattle

Because mouse preimplantation embryos grow and differentiate in the absence of exogenous factors, endogenous factors must sustain the embryo during the first six cleavage divisions (1). These early cleavage divisions serve two unique functions in mammals: (*a*) the generation of progenitors of the trophoblasts and extraembryonic membranes, and (*b*) the generation of the embryonic anlagen from the inner cell mass (ICM), and, hence, the embryo proper. Fate maps indicate that after implantation mammalian gastrulation and neurulation may be mechanistically and morphologically similar to that of nonmammalian vertebrates, such as *Xenopus*. However, unlike *Xenopus*, the unfertilized egg does not have partitioned cytosolic determinates, but must generate positional information during the six preimplantation cleavage divisions. Also, unlike the abbreviated synchronous cleavage cell cycles of *Xenopus* that precede gastrulation, mouse preimplantation embryos have near-normal cell cycle times (2–4) that may be regulated by growth factors. The paradigm for intercellular regulation of growth and differentiation is the interaction of growth factor ligands and receptors.

Until recently, only indirect evidence indicated that preimplantation em-

Acknowledgments: This work was supported by the Office of Health and Environmental Research, U.S. Department of Energy, Contract No. DE-AC03-76-SF01012, by the National Institutes of Health National Research Service Award 5-T32-ES07106 from the National Institute of Environmental Health Sciences, by NIH Grants HD-23539 and HD-23651, by Medical Research Council (Canada) Grant MT-4854, and by the Alberta Heritage Foundation for Medical Research.

Until recently, only indirect evidence indicated that preimplantation embryos make growth factors. First, cultured preimplantation embryos produce transforming growth factor-like bioactivity that promotes anchorage-independent growth (5). Second, shortly after implantation in the uterus, rodent embryos produce basic fibroblast growth factor (bFGF) (6–7), transforming growth factors α (8–10) and β (11) (TGFα and TGFβ), insulin-like growth factor II (IGF-II) (12–14), and int-2 (15), and human postimplantation embryos produce IGF-II transcripts (16–19). Preimplantation mouse embryos bind and respond to insulin (20–21). However, these factors have been implicated only in the later phases of postimplantation growth and differentiation, and their presence does not indicate whether or not these growth factors are produced by the preimplantation embryo. Other evidence for growth factor production in early mammalian embryogenesis comes from teratocarcinoma cells, which are thought to be similar to the primitive ectoderm (22). The differentiated progeny of some of the teratocarcinoma lines are also equivalent to endodermal cells derived from the blastocyst (22–23). Undifferentiated teratocarcinoma cells produce platelet-derived growth factor (PDGF) (24), and three stem cell polypeptide growth factors (25). Differentiated teratocarcinoma cells also respond to nerve growth factor (NGF) (26–28), IGF-II (29–30), epidermal growth factor (EGF) (23, 31), and PDGF (24–25). Whether these transformed cells accurately reflect the conditions in preimplantation embryos is not known because transformation may be caused by improper expression of growth factors or receptors in these lines. However, embryonic stem (ES) cells are derived from the ICM and may provide an avenue for studying expression and function of growth factor ligands and receptors in early mouse development.

Direct evidence for growth factor transcripts in low copy number in preimplantation embryos has been heretofore impossible to obtain. Localization of mRNA transcripts in embryos by in situ hybridization is difficult (32). Thousands of embryos are required to detect high-copy-number transcripts, such as histone or actin, by RNA blotting analysis (33–34). We have used mRNA phenotyping, a sensitive method for assaying unambiguously and simultaneously the accumulation of several growth factor transcripts in small numbers of mouse embryos (32, 35), embryonal carcinoma cells, and ES cells.

RESULTS

We developed a method for phenotyping mRNA in small numbers (1–100) of mouse embryos. The method consists of three linked techniques: a microadaptation of the guanidine thiocyanate/CsCl technique for isolating whole RNA (32), followed by reverse transcription (RT) with oligo(dT) or specific antisense oligonucleotide priming, and cDNA amplification by polymerase chain reaction (PCR). The products of the first-strand cDNA synthesis are divided and amplified separately by sequence-specific primers to produce a phenotype of growth factor transcripts. The method is highly sensitive: Messages from a single cell, a single embryo, or as few as 10 synthetic RNA transcripts can be detected (32, 36–37). The primers bracket a target sequence of diagnostic length of 0.2–0.5 kb and are chosen for (*a*) sequence

Table 1. Growth factor ligand mRNA transcripts found in preimplantation mouse embryos by RT-PCR.

Growth Factor	E	2	4	8	B	EC	Immunolocalization in Blastocyst
TGFα	+	–	+	+	++	+	ICM + TE
IGF-II	–	+	+	+	++	+	ICM + TE
kFGF	nd	nd	+	+	++	+	ICM
PDGF-A	+	–	+	+	++	+	ICM + TE
TGFβ_1	–	+	+	+	++	+	Some ICM + TE
IL-6	+	nd	nd	+	+	nd	nd

Note: E = unfertilized egg; 2, 4, and 8 = cell number in embryo; B = blastocyst; EC = undifferentiated Nulli cells; nd = not determined; TE = trophectoderm; ICM = inner cell mass.

specificity; (*b*) potential diagnostic traits, such as restriction endonuclease sites or cDNA inclusion; and (*c*) unique interaction with only the mature, processed transcript.

We found that preimplantation mouse embryos synthesize TGFα, TGFβ_1, PDGF-A, Kaposi's sarcoma-type fibroblast growth factor (kFGF) (38), interleukin-6 (IL-6), and IGF-II transcripts (Table 1), but do not synthesize insulin or IGF-I (Table 2). Blastocysts do not synthesize NGFβ, granulocyte colony-stimulating factor (G-CSF), bFGF, or EGF transcripts (Table 2). These transcripts fall into four temporal classes: (*a*) not transcribed at any time before implantation; (*b*) present as maternal transcripts, destroyed, and resynthesized as zygotic transcripts (TGFα, PDGF-A); or (*c*) transcribed only as zygotic transcripts (TGFβ_1, IGF-II) (Table 1). A fourth class of transcripts represented by β-actin and the metalloproteinase

Table 2. Growth factor ligand mRNA transcripts not found in preimplantation mouse embryos by RT-PCR.

Growth Factor	EC	2	4	8	B	Post Implantation
EGF	nd	nd	nd	nd	–	–
IGF-I	–	–	–	–	–	+
Insulin	–	–	–	–	–	–
bFGF	nd	nd	nd	nd	–	–
NGFβ	nd	nd	nd	nd	–	+
G-CSF	nd	nd	nd	nd	–	–

Note: EC = undifferentiated Nulli cells; 2, 4, and 8 = cell number in embryo; B = blastocyst; nd = not determined.

Table 3. Growth factor receptor mRNA transcripts found in preimplantation mouse embryos by RT-PCR.

Growth Factor Receptor	E	2	4	8	B	EC
IGF-I-R	–	–	–	+	+	+
IGF-II-R	–	+	+	+	+	+
Insulin-R	–	–	–	+	+	+
CSF-I-R	nd	+	+	+	+	–
PDGFα-R	nd	nd	nd	nd	+	+
PDGFβ-R	nd	nd	nd	+	+	+

Note: E = unfertilized egg; 2, 4, and 8 = cell number in embryo; B = blastocyst; EC = undifferentiated Nulli cells; R = receptor; nd = not determined.

scripts for IGF-I receptor, IGF-II receptor, and insulin receptor are present after the activation of the zygotic genome (Table 3).

Because not all growth factor transcripts are translated, we next asked whether a set of growth factor genes were expressed as polypeptides in mouse blastocysts (39). We found immunochemical evidence for the translation products of the three ligands TGFα, $TGF\beta_1$, PDGF (35), IGF-II, and kFGF (Table 1). Because IGF-I receptor, IGF-II receptor, and insulin receptor were transcribed (Table 3), but of the corresponding ligands only IGF-II ligand was transcribed, we tested for function of the three insulin family receptors. The growth factor-mediated stimulation of incorporation of radiolabeled amino acids into protein in cultured blastocysts was measured by modifying the technique of Harvey and Kaye (20). IGF-II, insulin (Fig. 1), and IGF-I (data not shown) stimulated significant increases

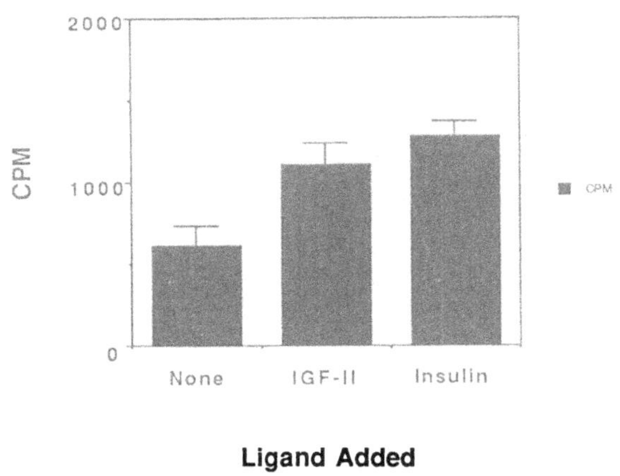

Fig. 1. Changes in incorporation of $[^{35}S]$methionine into cellular proteins of mouse blastocysts cultured with growth factors.

Table 4. Expression of growth factors and their receptors during differentiation of embryonic stem (ES) cells.

	Differentiated State (Relative Expression)			
Growth Factor or Receptor mRNA	Stem Cell	Embryoid Bodies	Cystic Bodies	Zoos
TGFβ$_1$	+	±	+	±
TGFβ2	–	–	±	+
CSF-I	±	–	–	+
CSF-I-R	–	–	–	+
IGF-I	+	+	+	+
IGF-II	±	±	+	+
Insulin-R	±	–	+	+
IGF-I-R	+	+	++	++
IGF-II-R	+	+	–	++
LIF/DIA	+	+	+	+
kFGF	+	–	±	–
NGF-R	++	++	++	++
EGF-R	–	nd	nd	++

Note: ES cells were differentiated as descrubed by Robertson in reference 74; R = receptor.

IGF-II, insulin (Fig. 1), and IGF-I (data not shown) stimulated significant increases in protein synthesis in cultured mouse blastocysts. These data suggest that insulin and IGF-I receptors are functioning, but it is currently not known if IGF-II receptors transduce signals (40). In addition, colony-stimulating factor-I (CSF-I) receptor, PDGFα receptor, and PDGFβ receptor transcripts were detected in blastocysts (Table 3). Preliminary results indicate that these receptors are functioning in preimplantation embryos (data not shown).

We have also studied the expression of growth factor ligands and transcripts in ES cells (Table 4). These cells express the phenotype of the ICM cells, but can contribute to all embryonic lineages and extraembryonic membranes when injected into the blastocyst (41). They provide a relatively easily manipulated culture model for early embryogenesis. We found that growth factor production in ES cells and blastocysts is similar but not identical.

DISCUSSION

The mouse embryo grows autonomously for only the first six divisions, at which time it interacts with, and implants in, the wall of the uterus. The controlling influences on these first divisions have not been studied. Zygotic gene transcription begins after the first cell division. To date, there has been no direct proof that preimplantation mouse embryos synthesize growth factors, although there are indirect results suggesting that embryos can bind and express specific growth

(23, 31), and peri-implantation mouse embryos cultured for 2 days produce TGF-like biological activity (5). We have found that TGFα, $TGFβ_1$, kFGF, PDGF-A, IL-6, and IGF-II genes are expressed in mouse blastocysts. There are recent reports that leukemia inhibitory factor/differentiation-inhibiting activity (LIF/DIA), IL-3, and $TGFβ_2$ are also synthesized by mouse blastocysts (42–44). This transcription is selective because blastocysts do not transcribe genes for EGF, bFGF, NGFβ, G-CSF, insulin, or IGF-I. It is likely that the TGFα, PDGF, and $TGFβ_1$ described here, individually or in combination, account for the TGF-like activity described by Rizzino (5). We have also found that preimplantation mouse embryos transcribe insulin receptor, IGF-I receptor, and IGF-II receptor mRNA and that these receptors may be functional in these embryos.

There is ample evidence for growth factor production and/or responsiveness in postimplantation fetal development. First, postimplantation rodent embryos at 7.5 days of gestation or later produce and/or respond to TGFβ (11), NGF (26), TGFα (8–9), bFGF (6–7), IGF-II (12, 30), and int-2, which has a sequence related to bFGF (15). Although most growth factors are not expressed in mouse embryos until the organogenesis phase (day 9), int-2 (15) and TGFα (9) are expressed by 7.5 days, only 2 days after implantation. Second, embryonal carcinoma stem cell lines, which resemble the pluripotential cells of the ICM in the preimplantation embryo, and their immediate progeny of endodermal lineage produce NGFβ (28), PDGF (24), bFGF (45), and TGFα (23). However, transformation of these stem cell lines may involve irregular expression of growth factor receptors or ligands. We found no NGFβ expression in blastocysts, whereas a trace of NGFβ has been found in F9 and PCC4 embryonal carcinoma cell lines (28). Third, embryonal carcinoma stem cells and/or their progeny specifically bind EGF (31), IGF-II (12, 30), PDGF (24), and NGF (26). Taken together, these findings suggest a role for growth factors in postimplantation embryonic growth development in mammals.

Growth factors have been implicated in the embryonic development of diverse nonmammalian species. bFGF and TGFβ appear to be morphogens for inducing mesoderm at the blastulation stage in *Xenopus* (46–49); TGFβ-like and EGF-like molecules may influence *Drosophila* development (50–52); EGF-like molecules may influence nematode development (53); and IGF-II-like and FGF-like molecules may influence chick development (54–55). These growth factors can induce differentiation, as in *Xenopus* (46–49), or induce both differentiation and mitosis, as in chick (54, 56); In the frog, growth factors may influence differentiation before the eighth cell division (3). In the mouse, maternal growth factor transcripts are replaced by zygotic growth factor transcripts before the sixth cell division. Early development in mouse has several other properties that distinguish it from that of frog. First, the egg is small, has little yolk, and quickly activates its zygotic transcription after fertilization (2). Second, the mouse has 12- to 24-h cell cycle times (2–3) after the first two cell cycles. These cycles have the normal $G_1/S/G_2/M$ periods, in contrast to the early cell divisions of frog that lack G_1 and G_2 (2, 4). The presence of G_1 and G_2 in cleavage-stage mouse embryos may allow transcription of growth factors, as well as the opportunity to be influenced by growth factors in a manner that modulates cell cycle.

manner that modulates cell cycle.

The accumulation patterns of growth factor transcripts in preimplantation mouse embryos fall into two classes. In one class, including PDGF-A and TGFα in mouse, maternal transcripts apparently disappear and are resynthesized in the zygote; frog FGF (48) and PDGF-A transcripts (57) act similarly. In the second class, transcripts survive the breakdown of maternal mRNA that is initiated during meiotic maturation, becomes quite dramatic in the 2-cell embryo, and continues up to the blastocyst stage (58). TGFβ_1 is an example of the second class. Similarly, in *Xenopus* the TGFβ-like Vg_1, which is localized to the vegetal hemisphere and may play a role in mesoderm induction, persists throughout early development (47). A comparable physiological role for TGFβ_1 or TGFβ_2 in mouse is not known.

What does the presence of these three early growth factor transcripts imply about their function in mouse embryos? We can separate growth factor functions by two criteria: direction and action. The direction can be within the embryo or between the embryo and the mother. The action can be to influence mitosis and/or differentiation. An intraembryonic premitogenic function is suggested by our findings that IGF-II ligand and its receptors, the IGF-I and IGF-II receptors, are expressed and that the IGF-I receptors and the IGF-II ligand are functional (59). An intraembryonic mitogenic function is suggested by the coincidental production by the autonomous blastocyst of the three growth factors (TGFα, TGFβ, and PDGF) belonging to a factor subset that sustains anchorage-independent growth (60). The onset of growth factor transcription from the zygotic genome in mouse roughly coincides with, or precedes, the differentiation of totipotent ICM cells into primitive ectoderm and endoderm. However, our evidence suggests that IGF-II acts to regulate anabolic, but not differentiation, processes. Several lines of evidence indicate that embryonic factors are directed at maternal tissue. The strongest evidence is the prolongation of corpus luteum lifespan in sheep by an embryonic protein thought to be ovine trophoblast protein-1 (61–62). This protein, which binds endometrial receptors and is the major translation product of ovine trophoblasts, was recently cloned and found to be highly homologous to a secreted polypeptide factor, α_{11}-interferon (62). Our evidence for embryonic-maternal communication is more circumstantial. First, TGFα and TGFβ_1 are known to be angiogenic (63–64), kFGF may be angiogenic (38), and the highest density of uterine capillary beds is opposite the implanting blastocyst (65). In addition, the uterine environment is hypoxic (66), a condition that promotes wound-healing cells to produce angiogenic factors (67). Finally, at the time of implantation, there is a surge of estrogen that increases EGF receptor expression in uterine glandular epithelium several-fold (68); TGFα is an EGF receptor-binding ligand. Taken together, these data indicate that embryonic growth factors may contribute to the induction of early angiogenesis and decidualization of the uterus. A possible maternal-to-embryonic influence is represented by the functional IGF-I and insulin receptor transcripts and proteins and the lack of IGF-I and insulin ligand transcripts and endogenous polypeptides. A second possible maternal-to-fetal interaction may be motivated by maternal CSF-I through the CSF-I receptor (69–70).

We speculate that IGF-II produced by mouse preimplantation embryos may modulate embryonic growth through IGF-I or IGF-II receptors, which are functional in early mouse embryos. The functional insulin and IGF-I receptors in mouse preimplantation embryos may transduce maternally derived signals because the embryo does not transcribe insulin or IGF-I mRNA. Our goals are to further dissect the pathways of IGF-II action to determine the relative functions of the IGF-I and IGF-II receptors. We also hope to determine if IGF binding proteins are present (71–73) and if they modulate IGF-II function in the preimplantation embryo.

REFERENCES

1. Biggers JD. New observations on the nutrition of the mammalian oocyte and the preimplantation embryo. In: Blandau RJ, ed. Biology of the blastocyst. Chicago: University of Chicago Press, 1971:319-27.
2. Pedersen RA. Potency, lineage, and allocation in preimplantation mouse embryos. In: Rossant J, Pedersen RA, eds. Experimental approaches to mammalian embryonic development. Cambridge: Cambridge University Press, 1986:3-33.
3. Molls M, Zamboglou N, Streffer C. A comparison of the cell kinetics of pre-implantation mouse embryos from two different mouse strains. Cell Tissue Kinet 1983;16: 277-83.
4. Graham CF, Morgan RW. Changes in the cell cycle during early amphibian development. Dev Biol 1966;14:439-60.
5. Rizzino A. Early mouse embryos produce and release factors with transforming growth factor activity. In Vitro Cell Dev Biol 1985;21:531-6.
6. Risau W. Developing brain produces an angiogenesis factor. Proc Natl Acad Sci USA 1986;83:3855-9.
7. Risau W, Ekblom P. Production of a heparin-binding angiogenesis factor by the embryonic kidney. J Cell Biol 1986;103:1101-7.
8. Lee DC, Rochford R, Todaro GJ, Villarreal LP. Developmental expression of rat transforming growth factor-α mRNA. Mol Cell Biol 1985;5:3644-6.
9. Twardzik DR. Differential expression of transforming growth factor-α during prenatal development of the mouse. Cancer Res 1985;45:5413-6.
10. Han VKM, Hunter ES III, Pratt RM, Zendegui JG, Lee DC. Expression of rat transforming growth factor alpha mRNA during development occurs predominantly in maternal decidua. Mol Cell Biol 1987;7:2335-43.
11. Heine U, Munoz EF, Flanders KC, et al. Role of transforming growth factor-β in the development of the mouse embryo. J Cell Biol 1987;105:2861-76.
12. D'Ercole AJ, Underwood LE. Ontogeny of somatomedin during development in the mouse. Dev Biol 1980;79:33-45.
13. Smith EP, Sadler TW, D'Ercole AJ. Somatomedins/insulin-like growth factors, their receptors and binding proteins are present during mouse embryogenesis. Development 1987;101:73-82.
14. Stylianopoulou F, Efstratiadis A, Herbert J, Pintar J. Pattern of the insulin-like growth factor II gene expression during rat embryogenesis. Development 1988; 103:497-506.
15. Jakobovits A, Shackleford GM, Varmus HE, Martin GR. Two proto-oncogenes implicated in mammary carcinogenesis, int-1 and int-2, are independently regulated

during mouse development. Proc Natl Acad Sci USA 1986;83:7806-10.
16. Beck F, Samani NJ, Penschow JD, Thorley B, Tregear GW, Coghlan JP. Histochemical localization of IGF-I and -II mRNA in the developing rat embryo. Development 1987;101:175-84.
17. Scott J, Cowell J, Robertson ME, et al. Insulin-like growth factor-II gene expression in Wilms' tumour and embryonic tissues. Nature 1985;317:260-2.
18. Ohlsson R, Holmgren L, Glaser A, Szpecht A, Pfeifer-Ohlsson S. Insulin-like growth factor 2 and short-range stimulatory loops in control of human placental growth. EMBO J 1989;8:1993-9.
19. Brice AL, Cheetham JE, Bolton VN, Hill NCW, Schofield PN. Temporal changes in the expression of the insulin-like growth factor II gene associated with tissue maturation in the human fetus. Development 1989;106:543-54.
20. Harvey MB, Kaye PL. Insulin stimulates protein synthesis in compacted mouse embryos. Endocrinology 1988;122:1182-4.
21. Mattson BA, Rosenblum IY, Smith RM, Heyner S. Autoradiographic evidence of insulin and insulin-like growth factor binding in early mouse embryos. Diabetes 1988; 37:585-9.
22. Martin GR, Evans MJ. Differentiation of clonal lines of teratocarcinoma cells: Formation of embryoid bodies in vitro. Proc Natl Acad Sci USA 1975;72:1441-5.
23. Adamson ED. Cell-lineage-specific gene expression in development. In: Rossant J, Pedersen RA, eds. Experimental approaches to mammalian embryonic development. Cambridge: Cambridge University Press, 1986:321-64.
24. Rizzino A, Bowen-Pope D. Production of PDGF-like factors by embryonal carcinoma cells and response to PDGF by endoderm-like cells. Dev Biol 1985;110:15-22.
25. Jakobovits A, Banda MJ, Martin GR. Embryonal carcinoma-derived growth factors: Specific growth-promoting and differentiation-inhibiting activities. In: Feramisco J, Ozanne B, Stiles C, eds. Growth factors and transformation. Cold Spring Harbor, NY: Cold Spring Harbor Laboratory, 1985:393-9.
26. Liesi P, Rechardt L, Wartiovaara J. Nerve growth factor induces adrenergic neuronal differentiation in F9 teratocarcinoma cells. Nature 1983;306:265-7.
27. Levi-Montalcini R, Booker B. Destruction of the sympathetic ganglia in mammals by an antiserum to a nerve-growth protein. Proc Natl Acad Sci USA 1960;46:384-91.
28. Dicou E, Houlgatte R, Brachet P. Synthesis and secretion of β-nerve growth factor by mouse teratocarcinoma cell lines. Exp Cell Res 1986;167:287-94.
29. Heath JK, Rees AR. Growth factors in mammalian embryogenesis. In: Growth factors in biology and medicine. Ciba Foundation Symp. 1985;116:3-22.
30. Heath JK, Shi W-K. Developmentally regulated expression of insulin-like growth factors by differentiated murine teratocarcinomas and extraembryonic mesoderm. J Embryol Exp Morph 1986;95:193-212.
31. Adamson ED, Hogan BLM. Expression of EGF receptor and transferrin by F9 and PC13 teratocarcinoma cells. Differentiation 1984;27:152-7.
32. Rappolee DA, Wang A, Mark D, Werb Z. Novel method for studying mRNA phenotypes in single or small numbers of cells. J Cell Biochem 1989;39:1-11.
33. Piko L, Clegg KB. Quantitative changes in total RNA, total poly(A), and ribosomes in early mouse embryos. Dev Biol 1982;89:362-78.
34. Giebelhaus DH, Heikkila JJ, Schultz GA. Changes in the quantity of histone and actin messenger RNA during the development of preimplantation mouse embryos.

Dev Biol 1983;98:148-54.
35. Rappolee DA, Brenner CA, Schultz R, Mark D, Werb Z. Developmental expression of PDGF, TGF-α, and TGF-β genes in preimplantation mouse embryos. Science 1988; 241:1823-5.
36. Rappolee DA, Mark D, Banda MJ, Werb Z. Wound macrophages express TGF-α and other growth factors in vivo: Analysis by mRNA phenotyping. Science 1988;241: 708-12.
37. Brenner CA, Adler RR, Rappolee DA, Pedersen RA, Werb Z. Genes for extracellular matrix-degrading metalloproteinases and their inhibitor, TIMP, are expressed during early mammalian development. Genes Dev 1989;3:848-59.
38. Delli Bovi P, Curatola AM, Kern FG, Greco A, Ittmann M, Basilico C. An oncogene isolated by transfection of Kaposi's sarcoma DNA encodes a growth factor that is a member of the FGF family. Cell 1987;50:729-37.
39. Assoian RK, Fleurdelys BE, Stevenson HC, et al. Expression and secretion of type β transforming growth factor by activated human macrophages. Proc Natl Acad Sci USA 1987;84:6020-4.
40. Roth R. Structure of the receptor for insulin-like growth factor II: The puzzle amplified. Science 1988;239:1269-71.
41. Hogan B, Costantini F, Lacy E. eds. Manipulating the mouse embryo; a laboratory manual. Cold Spring Harbor, NY: Cold Spring Harbor Laboratory, 1986.
42. Conquet F, Brûlet P. Developmental expression of leukaemia inhibitory factor (LIF) gene in preimplantation blastocyst and in extraembryonic tissue of the mouse embryo. J Cell Biochem 1990;14E(suppl):118.
43. Mummery CL, Feyen A, Freund E, Kruijer W, van den Eynden AJM. Expression of TGFβ, PDGF and IGF during the bipotential differentiation of embryonic stem cells. J Cell Biochem 1990;14E(suppl):61.
44. Murray R, Choy-pik C, Lee F. Hemopoietic growth factor expression in pre- and post-implantation mouse embryos. J Cell Biochem 1990;14E(suppl):93.
45. van Veggel JH, van Oostwaard TMJ, de Laat SW, van Zoelen EJJ. PC13 embryonal carcinoma cells produce a heparin-binding growth factor. Exp Cell Res 1987;169: 280-6.
46. Slack JMW, Darlington BG, Heath JK, Godsave SF. Mesoderm induction in early *Xenopus* embryos by heparin-binding growth factors. Nature 1987;326:197-200.
47. Weeks DL, Melton DA. A maternal mRNA localized to the vegetal hemisphere in *Xenopus* eggs codes for a growth factor related to TGF-β. Cell 1987;51:861-7.
48. Kimelman D, Kirschner M. Synergistic induction of mesoderm by FGF and TGF-β and the identification of an mRNA coding for FGF in the early *Xenopus* embryo. Cell 1987;51:869-77.
49. Rosa F, Roberts AB, Danielpour D, Dart LL, Sporn MB, David IB. Mesoderm induction in amphibians: The role of TGF-β2-like factors. Science 1988;239:783-5.
50. Padgett RW, St Johnston RD, Gelbart WM. A transcript from a *Drosophila* pattern gene predicts a protein homologous to the transforming growth factor-β family. Nature 1987;325:81-4.
51. Wharton KA, Johansen KM, Xu T, Artavanis-Tsakonas S. Nucleotide sequence from the neurogenic locus notch implies a gene product that shares homology with proteins containing EGF-like repeats. Cell 1985;43:567-81.
52. Hafen E, Basler K, Edstroem J-E, Rubin GM. Sevenless, a cell-specific homeotic gene of *Drosophila*, encodes a putative transmembrane receptor with a tyrosine kinase

domain. Science 1987;236:55-63.
53. Greenwald I. Lin-12, a nematode homeotic gene, is homologous to a set of mammalian proteins that includes epidermal growth factor. Cell 1985;43:583-90.
54. Bell KM. The preliminary characterization of mitogens secreted by embryonic chick wing bud tissues in vitro. J Embryol Exp Morph 1986;93:257-65.
55. Goldin GV, Opperman LA. Induction of supernumerary tracheal buds and the stimulation of DNA synthesis in the embryonic chick lung and trachea by epidermal growth factor. J Embryol Exp Morph 1980;60:235-43.
56. Engstrom W, Bell KM, Schofield PN. Expression of the insulin-like growth factor II gene in the developing chick limb. Cell Biol Int Rep 1987;11:415-21.
57. Mercola M, Melton DA, Stiles CD. Platelet-derived growth factor A chain is maternally encoded in *Xenopus* embryos. Science 1988;241:1223-5.
58. Schultz GA. Utilization of genetic information in the preimplantation mouse embryo. In: Rossant J, Pedersen, RA, eds. Experimental approaches to mammalian embryonic development. Cambridge: Cambridge University Press, 1986:239-65.
59. Rappolee DA, Schultz GA, Pedersen RA, Sturm K, Werb Z. An endogenous growth factor-receptor circuit in preimplantation mammalian development. J Cell Biochem 1989;13B(suppl):200.
60. Anzano M, Roberts AB, Sporn MB. Anchorage independent growth of primary rat embryo cells is induced by platelet derived growth factor and inhibited by type-beta transforming growth factor. J Cell Physiol 1986;126:312-6.
61. Weitlauf HM. Biology of implantation. In: Knobil E, Neill J, eds. Physiology of reproduction. New York: Raven Press, 1988:231-62.
62. Imakawa K, Anthony RV, Kazemi M, Marotti KR, Polites HG, Roberts RM. Interferon-like sequence of ovine trophoblast protein secreted by embryonic trophectoderm. Nature 1987;330:377-9.
63. Schreiber AB, Winkler ME, Derynck R. Transforming growth factor-α. A more potent angiogenic mediator than epidermal growth factor. Science 1986;232:1250.
64. Roberts AB, Sporn MB, Assoian RK, et al. Transforming growth factor type β: Rapid induction of fibrosis and angiogenesis in vivo and stimulation of collagen formation in vivo. Proc Natl Acad Sci USA 1986;83:4167-71.
65. Williams MF. The vascular architecture of the rat uterus is influenced by estrogen and progesterone. Am J Anat 1984;83:274-85.
66. Yochim JM. Intrauterine oxygen tension and metabolism of the endometrium during the preimplantation period. In: Blandau RJ, ed. The biology of the blastocyst. Chicago: University of Chicago Press, 1981:363-82.
67. Knighton DR, Hunt TK, Scheuenstuhl H, Halliday BJ, Werb Z, Banda MJ. Oxygen tension regulates the expression of angiogenesis factor by macrophages. Science 1983;221:1283-5.
68. Mukku VR, Stancel GM. Regulation of epidermal growth factor receptor by estrogen. J Biol Chem 1985;260:9820-4.
69. Pollard DW, Ladner MB, Stanley ER. Apparent role of the macrophage growth factor, CSF-1, in placental development. Nature 1987;330:481-6.
70. Regenstreif LJ, Rossant J. Expression of the c-fms proto-oncogene and of the cytokine, CSF-1, during mouse embryogenesis. Dev Biol 1989;133:284-94.
71. Baxter, RC, Martin, JL. Binding proteins for the insulin-like growth factors: Structure, regulation and function. Prog Growth Factor Res 1989;1:49-68.

72. Elgin RG, Busby WH Jr, Clemmons DR. An insulin-like growth factor (IGF) binding protein enhances the biologic response to IGF-I. Proc Natl Acad Sci USA 1987; 84:3254-8.
73. Sussenbach JS. The structure of the insulin-like growth factor family. Prog Growth Factor Res 1989;1:33-48.
74. Robertson EJ, ed. Teratocarcinomas and embryonic stem cells; a practical approach. Oxford: IRL Press, 1987.

16

Colony-Stimulating Factor I in the Mouse and Human Uteroplacental Unit

Jeffrey W. Pollard,[1] Serge Pampfer,[1] Eric Daiter,[2] David Barad,[2] and Robert J. Arceci[3]

[1]Department of Developmental Biology and Cancer and [2]Department of Obstetrics and Gynecology, Albert Einstein College of Medicine, Bronx, New York, and [3]Department of Hematology/Oncology, Dana Farber Cancer Institute, The Children's Hospital, Boston

The uterus has long been thought of as producing secretions (histotrophe) that support the development of the fetus (1). These secretions undoubtedly include polypeptide growth factors (2–3). Such growth factors may act to prepare the uterus for receipt of the blastocyst, to stimulate the proliferation and differentiation of the extraembryonic tissues, to influence the development of supporting structures, such as the maternal vasculature, or to act directly on the fetus. In addition, the placenta produces a variety of cytokines whose roles may be similar to those produced from the uterus (4). Interestingly, several of these uterine or placental growth factors were originally described to act on, or be produced by, hematopoietic cells (3). Thus, there may be an overlap between the growth factor regulation of embryonic development and hematopoiesis. In part, this overlap may be to regulate the maternal immunological response to the semi-allogenic fetus (5).

The number of growth factors reported to be synthesized by the uterus or placenta is expanding rapidly (see 2, 3, and 6 for reviews). Female sex steroids have been shown to elevate the uterine concentration of some of these cytokines (3). These data have to be interpreted with caution, however, since estrogens elicit considerable uterine edema (7) and alter the cellular composition of the uterus (8). Thus, in situ techniques are needed to confirm that the uterine source of the growth factor is due to synthesis by a resident cell type and is not due to the steroid-induced

Acknowledgments: This work was supported by grants from the National Institute of Health, HD-25074-01 (JWP), IGM-38156-01 (RJA) and the Albert Einstein Core Cancer Center Grant P30-CA-1330 (JWP). We thank Dr. E.R. Stanley for performing the human RIA and for continuous support and criticism of this research.

recruitment of cells able to synthesize specific growth factors or to the influx of serum growth factors into the uterus.

In this chapter we concentrate on the colony-stimulating factor I (CSF-I) and its receptor as an example of a growth factor whose uterine epithelial synthesis is under the regulation of female sex steroids (9) and whose receptor is expressed in both maternal and fetal uteroplacental tissue (10).

EXPRESSION OF CSF-I AND ITS RECEPTOR IN THE MOUSE UTEROPLACENTAL UNIT

CSF-I was originally purified away from other hematopoietic colony-stimulating activities as a growth factor specifically required for the survival, proliferation, and differentiation of mononuclear phagocytes (11). It is a homodimeric glycoprotein with a variable subunit molecular weight of between 14,000 and 21,000 due to C-terminal proteolysis (12). Glycosylation of these subunits results in a native protein with molecular weights ranging from 50 to 70,000 d (12). In the mouse, CSF-I is found in appreciable quantities in serum (approximately 440 pM) and is detected in all tissues tested (13). Its action is mediated through a single class of high-affinity receptors (CSF-I-R) (14–15). This receptor is a 185,000-d transmembrane tyrosine kinase shown to be the product of the c-*fms* proto-oncogene (16).

Early screens of tissues revealed the gravid uterus as one of the richest sources of colony-stimulating activity (17–18). This activity is largely, if not entirely, CSF-I (13). During gestation the uterine concentration of CSF-I is elevated almost 1000-fold. CSF-I is also found in the placenta at high, but constant concentrations and at much lower, but variable concentrations in the fetus (13). It was also present in the amniotic fluid, where it showed a peak in concentration of about 6-fold over the serum concentration at day 14 of pregnancy (19–20). CSF-I mRNA was confined to the luminal and glandular epithelium through the entire gestation period, as shown by in situ hybridization (9–10, 21). Its concentration was elevated at least 100-fold to reach a peak at day 14–16 of pregnancy (10). The uterine CSF-I mRNA was predominantly a 2.3-kb form with a minor 4.6-kb species. These different forms are the result of an alternative splice of exons 9 and 10 (22) (Pollard, unpublished observations) with the 2.3-kb form containing exon 9. The splicing out of exon 10 results in the loss of three repeats of the AUUUA motif known to confer instability on other lymphokine mRNAs (23). The detection of CSF-I mRNA in the epithelium, as well as immunohistochemical detection of CSF-I in the uterine epithelium (J. Hunt and J.W. Pollard, unpublished observations), indicate that the elevated concentrations of CSF-I in the uteroplacental unit during gestation results largely, if not entirely, from synthesis by the uterine epithelium (9) and is not the result of accumulation from the serum, where it has a very short half-life (24). Serum CSF-I concentrations, however, are also elevated during pregnancy, but only by 1.4-fold. This elevation correlates with an increase in splenic CFU-C and a monocytosis (13). Since there is such a large elevation of uterine CSF-I through gestation, this serum elevation may be the result of overflow from the uterus.

Elevated uterine concentrations of CSF-I could also be induced by female sex steroid hormones given to ovariectomized animals (9). Thus, either estradiol-17β or progesterone alone caused a small but significant elevation of CSF-I concentration over that detected in the ovariectomized control. Given together, however, in regimens that approximate the concentrations found in the early pregnancy, there was a synergistic effect on uterine CSF-I concentration as assessed after 6 days of treatment. If a decidual stimulus of arachis oil was given intraluminally to the uterus of these steroid hormone-treated animals, then a further elevation of uterine CSF-I concentration was induced. This increase in uterine CSF-I concentration correlated with an induction of the 2.3-kb form of the CSF-I mRNA. The concentration of the CSF-I mRNA and CSF-I obtained in this case were similar to those found at a comparable period of pregnancy (9–10). Removal of estrogen following administration of a decidual signal to the progesterone- and estradiol-17β-treated animals resulted in loss of the CSF-I induction and a return to concentrations below that of the control, untreated uterus. It can be concluded, therefore, that the uterine synthesis of CSF-I early in pregnancy is under the regulation of progesterone and estradiol-17β and a decidual stimulus. It is not certain however, whether the steroid hormones act directly to elevate CSF-I mRNA concentrations or act via an intermediary signal.

Decidualization per se is not required for uterine CSF-I induction since doses of estradiol-17β that are relatively inhibitory to this process still induce the elevation in uterine CSF-I concentration, providing that progesterone is present and a decidual stimulus occurs. The nature of this decidual stimulus is not known. The maximal elevation of CSF-I mRNA and CSF-I is greatest during the period of gestation that cannot be mimicked by the artificial induction of deciduomata in hormone-treated animals. After day 16 of pregnancy, the CSF-I mRNA concentration declines, even though progesterone and estradiol-17β concentrations are high. These data suggest that additional positive or negative regulators of uterine CSF-I biosynthesis may be defined.

The CSF-I-R displays a complex pattern of expression during pregnancy (10, 21). Initially, a high level of expression is detected in cells of the primary decidual zone (Fig. 1). These cells are unlikely to be macrophages since they are largely excluded from this area (25). Morphologically, the cells expressing the CSF-I-R appear to be large decidual cells (10). Coincidentally, there is a low but detectable level of CSF-I-R mRNA expression in the invading trophectoderm. In addition, the secondary giant cells extending away from the ectoplacental cone towards the disintegrating epithelium express very high levels of the CSF-I-R mRNA. As pregnancy progresses the expression of CSF-I-R mRNA declines in decidual cells such that when the placenta is mature, there is no expression in the decidua capularis and very-low-to-undetectable expression in the decidua basalis. During the formation of the placenta, however, the decidua basalis does have an appreciable level of CSF-I-R mRNA expression (10).

By day 9 of gestation, as the placenta is forming, there is significant expression of CSF-I-R mRNA in the spongiotrophoblastic region and a high level of expression in the trophoblastic giant cells. At day 14, when the placental layers are easily discernible, the levels of expression are the following: giant trophoblastic

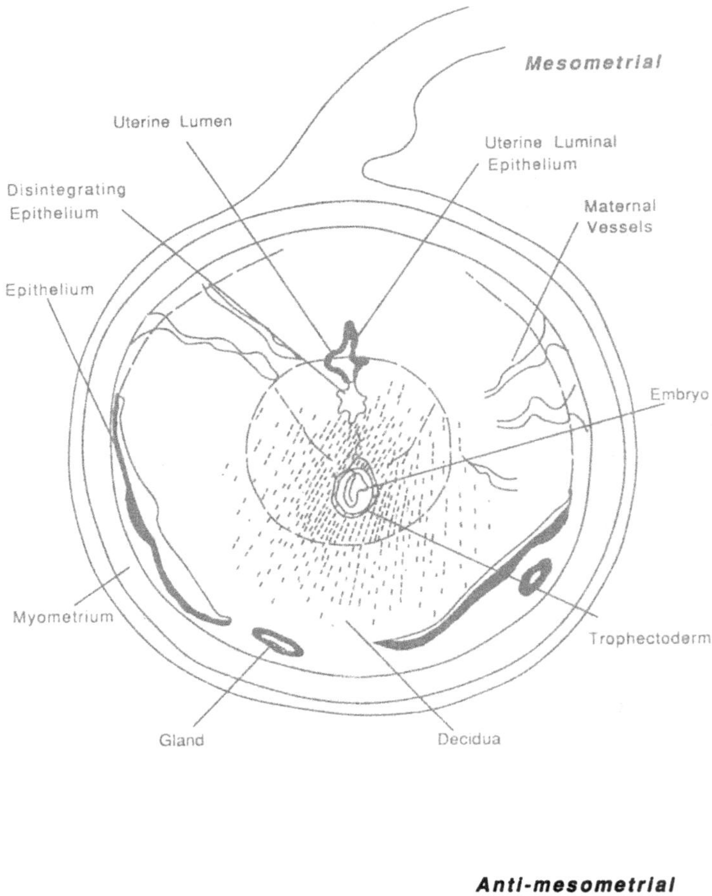

Fig. 1. A schematic depiction of a transverse section of a day-7 uterus at the peak of the decidual reaction indicating the areas of CSF-I and CSF-I-R mRNA expression. (Stipple = CSF-I-R nRNA; filled areas = CSF-I mRNA.)

layer > spongiotrophoblastic layer > labyrinthine layer. There is also expression on the visceral yolk sac (10, 21). This pattern, once established, persists until term and is summarized in Figure 2.

CSF-I IN THE HUMAN UTEROPLACENTAL UNIT

Using a radioimmunoassay specific for biologically active CSF-I, we have demonstrated CSF-I in human endometrium (12.8 ng/g tissue wet weight, n = 13), the placenta (11.67 ± 1.15-ng/g wet weight, n = 11), and amniotic fluid (3.84 ± 0.74

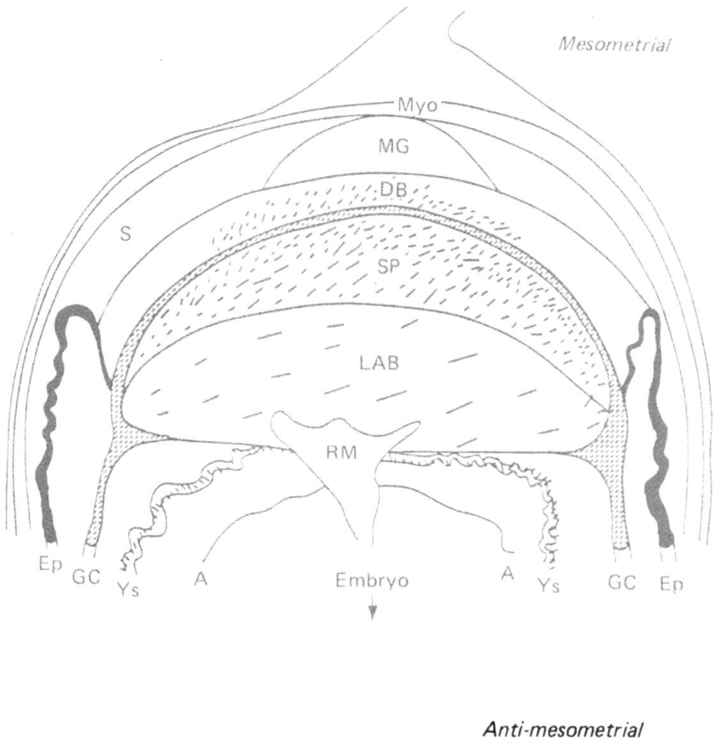

Fig. 2. Schematic diagram of a day-14 uteroplacental unit illustrating the expression of CSF-I and CSF-I-R mRNA. (Stipple = CSF-I-R mRNA; filled areas = CSF-I mRNA; Ep = uterine epithelium; S = stroma; GC = giant cell layer; Ys = visceral yolk sac; Sp = spongiotrophoblastic layer; Lab = labyrinthine layer; A = amnion; FM = fetal mesenchyme; Myo = myometrium; MG = metrial gland; DB = decidua basalis.)

ng/mL, n = 5) during the first trimester of gestation. The presence of CSF-I in amniotic fluid at similar concentrations during the first trimester has also been reported (26). The CSF-I concentrations reported in third-trimester (26) endometrium were 3-fold higher than those presented here for the first-trimester endometrium. Uterine synthesis of this CSF-I is suggested by the presence of a 4.0-kb mRNA in first-trimester decidual tissue (Fig. 3*a*). This mRNA concentration was elevated compared to that detected in nonpregnant endometrium (Fig. 3*a*). Similar preliminary observations have also been reported (27). A 4.0-kb CSF-I mRNA was detected in placenta from first trimester (Fig. 3*a*). A similar-sized mRNA has also been reported in an undated placental sample (28). In situ hybridization will

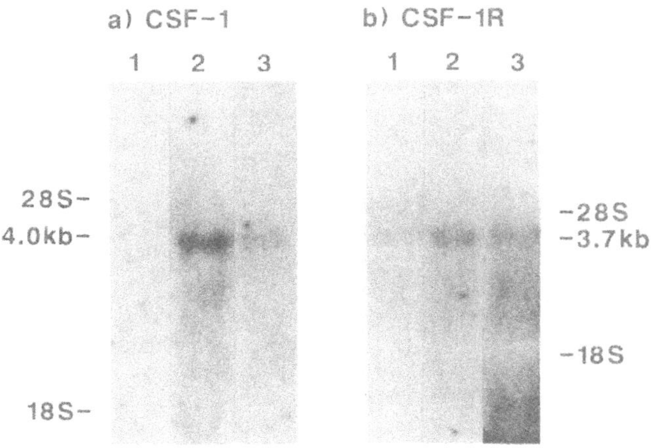

Fig. 3. Expression of CSF-I and CSF-I-R mRNA in human uteroplacental tissue. A composite autoradiograph of Northern blots of RNA probed with radiolabeled CSF-I (*a*) or CSF-I-R cDNA (*b*). *a* and *b*: Lane 1 = proliferative-phase endometrium; lane 2 = first-trimester endometrium; lane 3 = first-trimester placenta. The positions of the 4.0-kb CSF-I mRNA (*a*) and 3.7-kb CSF-I-R mRNA (*b*) are indicated. 28S and 18S mark the position of the mRNA species.

confirm if the placental CSF-I mRNA is due to uterine contamination of placental tissues or if there is a placental source of CSF-I.

We have shown the expression of a 3.7-kb CSF-I-R mRNA in both proliferative-phase and first-trimester endometrium (Fig. 3*b*). This expression may be due either to the presence of macrophages that are known to be present in human endometrium and that express the CSF-I-R, or to specific expression in endometrial cells. CSF-I-R mRNA has been detected in human placenta (29). We have confirmed these observations by demonstrating CSF-I-R mRNA expression in first-trimester placenta (Fig. 3*b*). In situ hybridization has located CSF-I-R mRNA in human cytotrophoblasts (30) and in the syncytial trophoblastic layer surrounding choriocarcinoma tissues (31). Expression of CSF-I-R mRNA has also been reported in third-trimester amnion and chorion (29). Therefore, the pattern of expression of the CSF-I-R mRNA in humans during pregnancy appears to be at least as complex as that described in the mouse.

CSF-I-R mRNA isolated from placenta contains an untranslated exon at the most 5' end that is not found in macrophages. The placental cell type utilizing this alternative exon remains to be identified. The alternative usage of this upstream exon allows transcription to begin from a placental-specific promoter (32). This suggests a unique regulation of CSF-I-R gene transcription in the placenta. Thus, each individual cell type that expresses the CSF-I-R may, by the use of alternative mRNA splicing, exploit different promoters to effect cell-specific regulation of the CSF-I-R gene during pregnancy.

Human choriocarcinoma cell lines express the CSF-I-R. Autophosphorylation of this receptor has been demonstrated in an immune-complex-kinase assay (33). This suggests that the placental CSF-I-R is a functional tyrosine kinase. The choriocarcinoma cell lines—JEG, JAR, and BEWO—produce CSF-I and may display autocrine growth (5). Interestingly, endometrial adenocarcinomas and ovarian carcinomas coexpress CSF-I and its receptor. The level of expression of these genes within the tumor and the serum concentration of CSF-I is correlated with, and may be predictive for, the progression of the disease (34–35). Thus, these neoplasms may also display CSF-I autocrine regulation of growth. Therefore, studies on the regulation of the CSF-I and CSF-I-R genes during normal pregnancy may contribute to an understanding of the genesis of these reproductive tract tumors.

WHAT ARE THE FUNCTIONS OF CSF-I IN PREGNANCY?

CSF-I is present in the uterus from the preimplantation period to parturition, suggesting that CSF-I may have a role throughout pregnancy. This role will vary according to the time and cell type examined. For example, CSF-I regulates the survival, proliferation, and differentiation of mononuclear phagocytes (11). It is also chemotatic for these cells (36). Macrophages accumulate in the uterus from early gestation (37). The high concentration of uterine CSF-I is probably responsible for this recruitment. CSF-I regulates the production by macrophages of many biologically active components, including cytokines (12). Thus, macrophages may affect the behavior of cells in the uteroplacental unit and, given their immunological role, even act in an immunoregulatory capacity by releasing immunosupressor molecules (38). In fact, a cytokine dialogue occurring within the uteroplacental unit has been proposed, with macrophages acting as the central cell receiving and sending messages (6).

CSF-I mRNA is elevated prior to implantation (10) and significantly elevated levels of CSF-I can be detected at implantation (13). These observations led us to suggest that CSF-I, in its transmembrane form may play a role in the implantation reaction (9) by interacting with receptors on the trophectoderm of the blastocyst. Little data pertain to this hypothesis. Recently, however, Tartakofsky (39) has shown that administration of CSF-I during the preimplantation period significantly reduces the success of implantation and increases the number of fetal resorptions. The most likely interpretation of these data, therefore, is that unphysiological doses of CSF-I given prematurely to blastocysts could result in inappropriate activation of the CSF-I-R tyrosine kinase, down-regulation of the CSF-I-R, and, consequently, a failure to respond appropriately to physiological concentrations of soluble, and perhaps transmembrane, CSF-I. Future studies will no doubt elucidate the role of CSF-I in embryo development.

The coincidence of maximal uterine CSF-I synthesis with both placental growth and CSF-I-R mRNA expression in decidua and trophoblasts, as well as the evolutionary conservation of the pattern of expression between mouse and humans,

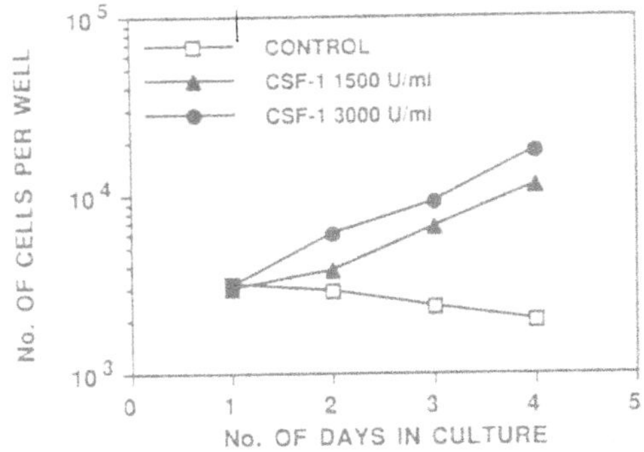

Fig. 4. Proliferation of placental cells in response to CSF-I. Day-14 mouse placenta were dissected free from the uterus and fetal tissues and membranes, chopped into small pieces, and placed in culture in α-MEM supplemented with 15% fetal calf serum (growth medium) and 1.32-nM human recombinant CSF-I. When outgrowth cells were confluent, they were trypsinized and seeded into 24-well Limbro dishes at 5×10^3 cells per well in growth medium either lacking CSF-I or containing it at the indicated concentrations (1 U = 0.44 fmoles). Cell numbers were assessed in triplicate daily.

suggests that CSF-I is involved in the regulation of placental growth and differentiation. The evidence that CSF-I is a placental growth factor is still slight. In mouse placental primary cell culture and in placentally derived cell lines, CSF-I stimulates cell proliferation (40). We have demonstrated a reproducible, dose-dependent effect on the proliferation of placental cells in culture (Fig. 4). Unfortunately, in neither experiment have the responding cells been unequivocally identified as trophoblasts.

Furthermore, it has not been directly established in the primary placental cell cultures that the responding cells proliferate as a result of stimulation by CSF-I or through the intermediary of a second cell type. In our experiment the cell types responding to CSF-I display cell-surface CSF-I receptors, as detected by immunofluorescence using a goat anti-CSF-I-R antibody. This suggests a direct effect of CSF-I on these cells. Substantially more evidence, however, will be required before CSF-I is firmly established as a regulator of trophoblast proliferation.

The highest level of CSF-I-R mRNA expression is on giant trophoblasts. This observation and the fact that uterine and placental CSF-I remains high even after the placenta has ceased to show substantial proliferation indicate that CSF-I may have other roles than acting as a regulator of proliferation. We suggested that one of these roles may be to regulate the production of placental hormones (20). Data have now accumulated, although much of it in preliminary form, that CSF-I regulates hCG and hPL production in cultured human-term cytotrophoblasts (5). Although this has not yet been shown to be a direct effect of CSF-I acting through its receptor, if these observations are correct, a classical hormone regulatory circuit appears to be in operation. Thus, progesterone synthesized by the corpus luteum allows pregnancy to continue and stimulate the synthesis of CSF-I by uterine epithelial cells. This CSF-I may act on trophoblasts, causing them to synthesize placental lactogens or, in humans, hCG. These hormones return to the ovary to maintain the corpus lutea and the synthesis of progesterone and, therefore, the continuation of pregnancy. This regulatory circuit has the novel feature that one link is a locally produced polypeptide growth factor.

SUMMARY

In summary, the concentration of uterine CSF-I is elevated throughout pregnancy. Even at later stages, in the mouse, when the placenta has differentiated and largely stopped proliferating, CSF-I concentrations remain very high. These data and the preliminary experiments described above suggest that CSF-I may have varying roles at different stages of pregnancy. First, it may play a role in implantation and the subsequent invasion of the trophectoderm into the uterine wall. Second, it may act as a trophoblastic growth factor. Third, it may regulate the endocrine functions of the decidua and placenta, and fourth, it may regulate the survival of trophoblasts. It may also play a role in the immunoregulation of the maternal response to the fetus. The osteopetrotic (op/op) mouse is entirely deficient in systemic CSF-I. Evidence indicates that the mutation lies within the CSF-I gene (41). This mutant mouse, therefore, should allow the delineation of the role of CSF-I in pregnancy.

REFERENCES

1. Corner GW. Cyclic changes in the ovaries and uterus of the sow and their relation to the mechanism of implantation. Cont Embryol Carn Instn 1921;13:119-46.
2. Brigstock DR, Heap RB, Brown KD. Polypeptide growth factors in uterine tissues and secretions. J Reprod Fertil 1989;85:747-58.
3. Pollard JW. Regulation of polypeptide growth factor synthesis and growth-factor-related gene expression in the rat and mouse uterus before and after implantation. J Reprod Fertil 1990;88:721-31.
4. Ohlsson R. Growth factors proto-oncogenes and human placental development. Cell Differ Dev 1989;28:1-16.
5. Wegmann TG. The cytokine basis for crosstalk between the maternal immune and reproductive systems. Curr Opinions Immunol 1990 (in press).
6. Hunt JS. Cytokine networks in the uteroplacental unit: Macrophages as pivotal regulatory cells J Reprod Immunol 1989;16:1-17.
7. Szego CM, Roberts S. Steroid action and interaction in uterine metabolism. Recent Prog Horm Res 1953;8:419-60.
8. Tchernitchin AN. Eosinophil-mediated non-genomic parameters of estrogen stimulation—a separate group of responses mediated by an independent mechanism. J Steroid Biochem 1983;19:95-100.
9. Pollard JW, Bartocci A, Arceci R, Orlofsky A, Ladner MB, Stanley ER. Apparent role of the macrophage growth factor, CSF-1 in placental development. Nature 1987; 330:484-6.
10. Arceci RJ, Shanahan F, Stanley ER, Pollard JW. The temporal expression and

location of colony stimulating factor-1 (CSF-1) and its receptor in the female reproductive tract are consistent with CSF-1 regulated placental development. Proc Natl Acad Sci 1989;86:8818-22.

11. Stanley ER, Guilbert LT, Tushinski RJ, Bartelmez SH. CSF-1-A mononuclear phagocyte lineage-specific hematopoietic growth factor. J Cell Biochem 1983;21: 151-9.
12. Sherr CJ, Stanley ER. Colony stimulating factor-1. In: Sporn MB, Roberts AB, eds. Peptide growth factors and their receptors. New York: Springer-Verlag, 1990 (in press).
13. Bartocci A, Pollard JW, Stanley ER. Regulation of colony stimulating factor-1 during pregnancy. J Exp Med 1986;164:956-61.
14. Guilbert LJ, Stanley ER. Specific interaction of murine colony-stimulating factor with mononuclear phagocytic cells J Cell Biol 1986;85:153-9.
15. Guilbert LJ, Stanley ER. The interaction of ^{125}I-CSF-1 with bone marrow derived macrophages. J Biol Chem 1986;261:4024-32.
16. Sherr CJ, Rettenmier CW, Sacca R, Roussel MF, Look AT, Stanley ER. The c-fms proto-oncogene product is related to the receptor for the mononuclear phagocyte growth factor CSF-1. Cell 1985;41:665-76.
17. Bradley TR, Stanley ER, Sumner MA. Factors from mouse tissues stimulating colony growth of mouse bone marrow cells in vitro. Aust J Exp Biol Med Sci 1971;49:595-603.
18. Rosendaal M. Colony-stimulating factor (CSF) in the uterus of the pregnant mouse. J Cell Sci 1975;19:411-23.
19. Azoulay M, Webb CG, Sachs L. Control of hematopoietic cell growth regulators during mouse fetal development. Mol Cell Biol 1987;7:3361-3.
20. Pollard JW, Arceci R, Bartocci A, Stanley ER. Colony stimulating factor-1: A growth factor for trophoblasts? In: Wegmann TG, Gill T III, Nisbet-Brown E, eds. The molecular and cellular immunobioloby of the maternal fetal interface. Oxford University Press, 1990 (in press).
21. Regenstreif LJ, Rossant J. Expression of the c-fms proto-oncogene and of the cytokine CSF-1 during mouse embryogenesis. Develop Biol 1989;133:284-94.
22. Ladner MB, Martin GA, Noble JA, et al. cDNA cloning and expression of murine macrophage colony stimulating factor from L929 cells. Proc Natl Acad Sci 1988; 85:6706-10.
23. Shaw G, Kamen R. A conserved AU sequence from the 3' untranslated region of GM-CSF mRNA mediates selective degradation. Cell 1986;46:659-67.
24. Bartocci A, Mastrogiannis DS, Magliorati G, Stockert RJ, Wolkoff AW, Stanley ER. Macrophages specifically regulate the concentration of their own growth factor in the circulation Proc Natl Acad Sci USA 1987;84:6179-83.
25. Tachi C, Tachi S. Role of macrophages in the maternal recognition of pregnancy. J Reprod Fertil 1989;37:63-8.
26. Ringler GE, Coutifaris C, Strauss JF, Allen JI, Geier M. Accumulation of colony stimulating factor 1 in amniotic fluid during human pregnancy. Am J Obstet Gynecol 1989;160:655-6.
27. Kauma S. (1989) Colony stimulating factor-1 (CSF-1) mRNA expression in human endometrium during the menstrual cycle and early pregnancy [Abstract]. Fertil Steril 1989;52:S43.
28. Wong GG, Temple PA, Leary AC, et al. Human CSF-1: Molecular cloning and expression of 4-kb cDNA encoding the urinary protein. Science 1987;235:1504-8.

29. Muller R, Tremblay JM, Adamson ED, Verma IM. Tissue and cell type-specific expression of two human c-onc genes. Nature 1983;304:454-6.
30. Hoshina M, Nishio A, Bo M, Boime I, Mochizuki M. The expression of the oncogene fms in human chorionic tissue. Acta Obstet Gynecol Japan 1985;37:2791-8.
31. Stuart SG, Simister NE, Clarkson SB, Kacinski BM, Shapiro M, Mellman I. Human IgG receptor (hFcR11; CD32) exists as multiple isoforms in macrophages, lymphocytes and IgG-transporting placental epithelium. EMBO J 1989;8:3657-66.
32. Visvader J, Verma IM. Differential transcription of exon 1 of the human c-fms gene in placental trophoblasts and monocytes. Mol Cell Biol 1989;9:1336-41.
33. Rettenmier CW, Sacca R, Furman WL, et al. Expression of the human c-fms protooncogene product (colony stimulating factor-1 receptor) on peripheral blood mononuclear cells and choriocarcinoma cell lines. J Clin Invest 1986;77:1740-6.
34. Kacinski BM, Carter D, Mittal K, et al. High level expression of fms proto-oncogene mRNA is observed in clinically aggressive human endometrial adenocarcinomas. Int J Radiat Oncol Biol Phys 1988;15:823-9.
35. Kacinski BM, Stanley ER, Carter D, et al. Circulating levels of CSF-1 (M-CSF), a lymphohematopoietic cytokine, may be a useful marker of disease status in patients with malignant ovarian neoplasm. Int J Radiat Oncol Biol Phys 1989;17:159-64.
36. Wang JM, Griffin JD, Rambaldi A, Chen ZG, Mantovani A. Induction of monocyte migration by recombinant macrophage colony-stimulating factor. J Immunol 1988;141:575-9.
37. Hunt JS, Manning LS, Mitchell D, Selanders JR, Wood GW. Localization and characterization of macrophages in murine uterus. J Leuk Biol 1985;38:255-65.
38. Hunt JS, Manning LS, Wood GW. Macrophages in murine uterus are immunosuppressive. Cell Immunol 1984;85:499-510.
39. Tartakovsky B. CSF-1 induces resorption of embryos in mice. Immunol Lett 1989; 23;65-70.
40. Athanassakis I, Bleackley C, Paetkau V, Guilbert L, Barr PJ, Wegmann TG. The immunostimulatory effect of T cells and T cell lymphokines on murine fetally derived placental cells. J Immunol 1987;138:37-41.
41. Wiktor-Jedrzejczak W, Bartocci A, Ferrante AW Jr, et al. Proc Natl Acad Sci USA 1990;89 (in press).

17

Autocrine-Paracrine Role of Lymphohematopoietic Cytokines at the Maternal-Fetal Interface

Thomas G. Wegmann and Larry Guilbert

Department of Immunology, University of Alberta, Edmonton

There are two facts concerning the interaction between the immune and reproductive systems that must be dealt with in any putative explanation of how this interaction occurs. The first is that there appears to be no requirement for an intact maternal immune system in order to obtain successful reproduction. The second is that immune interaction can both positively and negatively affect reproductive outcome. The first conclusion arises out of the fact that doubly mutant scid beige mice can be bred under germ-free conditions. These mice, who completely lack normal B- and T-cell function and most NK cell function have been reported to produce normal fetuses (1). The latter conclusion is best illustrated by the fact that under a variety of circumstances, spontaneous fetal resorption can be prevented by immunizing with cells bearing paternal H-2 haplotype (2–3). In other circumstances immunizing female mice with certain tumor cells leads to increased spontaneous abortion (4).

This type of observation prompted us to postulate that the immune system interacts with the reproductive system through growth-promoting lymphohematopoietic cytokines, a concept we have termed *immunotrophism*. We initially observed that members of the CSF family of cytokines, including CSF-I, GM-CSF, and IL-3 can lead to the proliferation of placental cells with trophoblastic characteristics in vitro (5). Following upon these observations, Armstrong and Chaouat confirmed that GM-CSF can stimulate the proliferation of readily identifiable trophoblast cells in vitro, with a maximal effect observed in pure ectoplacental cone trophoblast derived from 7.5-day mouse embryos (6). Additional support for an active role of immune cell-derived GM-CSF came from our demonstration that GM-CSF is released from decidual cell cultures. This release is impaired if one removes maternal T-cells from the mother during gestation by injecting anti-T-cell monoclonal antibodies prior to assessment (7).

Such maternal T-cell removal leads to other effects as well, depending on the strain combination. It invariably leads to decreased placental proliferation and phagocytosis of latex beads in the strain combinations we have studied (6). In some instances, but not in others, it leads to increased fetal resorption (8). In accord with immunotrophic stimulation of placental growth are observations in MRL mice that exhibit a spontaneous lupus-like syndrome involving hyper T-cell reactivity and global autoimmunity. During gestation these mice exhibit very large placentas whose size can be reduced to normal by maternal T-cell removal during pregnancy. Parallel observations on placental phagocytosis give comparable results. Injection of these autoimmune spleen cells into mice undergoing spontaneous abortion adoptively transfers resistance to the abortion as well as increased placental and fetal size when compared to injection of sister-strain cells as a control (8). Finally, we have shown recently that injection of GM-CSF as well as IL-3 into mice prone to elevated spontaneous abortion can significantly reduce the level of spontaneous resorption (7, 9). Thus, there are a number of reasons for believing that cytokine cross talk is occurring between the immune and reproductive systems.

It should be noted in passing that recent experiments indicate that activated NK cells are capable of causing spontaneous abortion. Initially, Baines and his colleagues observed a correlation between the number of resorbing fetal-placental units and increased appearance of asialo-GM_1-positive cells in the vicinity (10). This led to experiments in which anti-asialo-GM_1 antibody was used to prevent spontaneous abortion (11). More recently, it has been reported that activated NK cells can adoptively transfer spontaneous abortion, but not if the cell preparations are treated with anti-asialo-GM_1 antibody (12). This effect can be mimicked by injecting the animals with either γ-interferon or TNFα (13). Thus, both positive and negative effects on spontaneous fetal resorption have been attributed to lymphohematopoietic cytokines, providing a basis for beginning to understand the molecular communication between the immune system and the reproductive system.

Our recent unpublished experiments utilizing human placental cells and choriocarcinomas in vitro raise the possibility that GM-CSF and CSF-I may be autocrine within the reproductive tissues themselves (14). Thus, the three human choriocarcinoma cell lines JAG, JAR, and BEWO release GM-CSF and CSF-I spontaneously in culture. If antibody to GM-CSF is added to the culture, proliferation of these cell lines is partially inhibited. The same is true upon addition of neutralizing antibody against the receptor for CSF-I. Term human-placental cultures transiently release GM-CSF as well as CSF-I. Both GM-CSF and CSF-I added to these cultures stimulate syncytialization of cytotrophoblasts and lead to an increased release of human chorionic gonadotropin and human placental lactogen. Shiverick and her colleagues have recently found that murine GM-CSF added to rat placental cultures leads to an increase in production of rat placental lactogen, thus providing corroboration of these results in a different species (15). None of this evidence constitutes convincing proof that these cytokines are autocrine for trophoblast cells within the placenta. This will require in situ hybridization studies, which are now in progress. In the meantime, it is our working hypothesis that the reason that one observes an effective immune system-reproductive system interaction is because the same cytokines are used by both systems in an autocrine-paracrine manner, and that these shared cytokines induce essential endocrine and other functional changes during placental and embryonic development. Some of the best evidence supporting this point of view has been developed by Stanley and his colleagues, which is reviewed in another part of this volume, and thus is not covered here.

REFERENCES

1. Croy AA, Chapeau C. Evaluation of the pregnancy immunotrophism hypothesis by assessment of the reproductive performance of young, adult mice of genotype scid/scid.bg/bg. J Reprod Fertil 1990 (in press).
2. Gill TJ III, Wegmann TG, eds. Immunoregulation and fetal survival. New York: Oxford University Press, 1987.
3. Beard RW, Sharp F, eds. Early pregnancy loss: Mechanisms and treatment. Ashton-under-Lyne, Lancs., UK: Peacock Press, 1988.
4. Tartakovsky, B. CSF-1 induces resorption of embryos in mice. Immunol Lett 1989; 23:65-70.
5. Athanassakis I, Bleackley RC, Paetkau V, Guilbert L, Barr PJ, Wegmann TG. The immunostimulatory effects of T-cells and T-cell lymphokines on murine fetally derived placental cells. J Immunol 1987;138:37.
6. Armstrong D, Chaouat G. Effects of lymphokines and immune complexes on murine placental cell growth in vitro. Biol Reprod 1989;38:400-6.
7. Wegmann TG, Athanassakis I, Guilbert L, et al. The role M-CSF and GM-CSF in fostering placental growth, fetal growth, and fetal survival. Transplant Proc 1989; 21:566.
8. Chaouat G, Menu E, Athanassakis I, Wegmann TG. Maternal T cells regulate placental size and fetal survival. Regional Immunol 1988;1:143-8.
9. Chaouat G, Menu E, Clark DA, Dy M, Minkowski M, Wegmann TG. Control of fetal survival in CBA × DBA/2 mice by lymphokine therapy. J Reprod Fertil 1990 (in press).
10. Gendron R, Baines MG. Infiltrating decidual natural killer cells are associated with spontaneous abortion in mice. Cell Immunol 1988;113:261-7.
11. De Fougerolles R, Baines MG. Modulation of the natural killer cell activity in pregnant mice alters the spontaneous abortion rate. J Reprod Immunol 1987;11: 147-53.
12. Kinsky R, Delage G, Rosin N, Ming NT, Hoffmann M, Chaouat G. A murine model of NK mediated resorption. Biol Reprod 1990 (in press).
13. Chaouat G, Menu E, Clark DA, Dy M, Minkowski M, Wegmann TG. Control of fetal survival in CBA × DBA/2 mice by lymphokine therapy. J Reprod Fertil 1990 (in press).
14. Guilbert L, et al. 1990 (manuscript in preparation).
15. Shiverick K. 1989 (personal communication).

18

Trophoblast Interferons

R. Michael Roberts, James C. Cross, Charlotte E. Farin, Peter W. Farin, Kyle K. Kramer, Harriet Francis, Clifford Librach, and Susan J. Fisher**

*Department of Animal Sciences, University of Missouri, Columbia, and *Department of Anatomy, University of California, San Francisco*

Interferons (IFN) are cytokines with complex effects on cells of the immune system. Ovine and bovine conceptuses produce IFN (originally designated oTP-1 and bTP-1) as their major secretory products during the peri-implantation period. They have been identified as IFNα, based on cDNA sequence analysis, N-terminal amino acid sequencing, and possession of characteristic antiviral and antiproliferative properties. They also inhibit PHA-induced lymphocyte blastogenesis without reducing interleukin-2 production. However, oTP-1 and bTP-1 are not typical 166-amino acid IFN, but belong instead to a less-studied IFNα-II subfamily whose members are 172 residues in length. Both are localized to trophectoderm during a limited period of conceptus development and appear to play an important antiluteolytic role in maintaining corpus luteum function in early pregnancy. IFN have also been reported to be associated with conceptus and placental tissues in the human, mouse, hamster, and pig. However, we have shown that the antiviral activity produced by the murine conceptus is not due to a typical IFNα-I or IFNβ. We have also failed to identify mRNA related to IFNα in the early porcine conceptus and found no evidence for production of IFN by first-trimester human cytotrophoblast cells or by cultured horse conceptuses collected at the time of maternal recognition of pregnancy in the mare. Therefore, although there is strong evidence that IFN has an important role in early pregnancy in cattle and sheep, this function may not be extended to nonruminant species.

Acknowledgments: This work was supported by grants from NIH (HD-21896) and CIBA-GEIGY, Basel, Ltd. We thank Gail Foristal for typing the manuscript and Dr. G.R. Adolph, Ernest-Boehringer Institute, Vienna, for assaying interferon activity in horse conceptus cultures. This is paper number 11,142 of the University of Missouri Agricultural Experiment Station.

CONTROL OF CORPUS LUTEUM FUNCTION BY THE CONCEPTUS

The conceptus communicates with the mother throughout pregnancy to initiate the adjustments necessary to provide for continued embryonic growth and development. The very earliest signals that alter maternal physiology are produced by the preimplantation embryo while it is still resident in the oviduct (1–2). However, it has long been evident that embryo transfer to nonpregnant recipients in a variety of species can be carried out well after the effects of these early "factors" are first manifested in the natural mother. It seems reasonable to conclude, therefore, that the responses induced by these factors in the mother are not necessary for successful establishment of pregnancy. In ewes, embryos can be successfully transferred as late as day 12–13 of the 17-day estrous cycle (3); while in cattle, where cycle length is 21 days, pregnancies have resulted from day-15 transfers (4).

By about 1980 it had become clear that the factors responsible for prevention of luteolysis in these two ruminant species were proteins secreted by the conceptus (5). Godkin, et al. (6) were able to identify the most likely candidate for this role in sheep by culturing conceptuses in vitro in the presence of radioactive amino acids and by analyzing the proteins released into the medium by two-dimensional electrophoresis. A protein consisting of a cluster of 3–4 isoforms of identical size (Mr ~18,000), but differing slightly in isoelectric point (pI 5.3–5.8) was identified as the major secretory product, appearing around day 13 and disappearing by day 23. It was easily purified and an antiserum prepared, and the protein, now named ovine trophoblast protein-1 (oTP-1), was shown to be produced by the trophectoderm (7). When purified oTP-1 was infused into the uteri of nonpregnant ewes, the interestrous interval was extended (8). Thus, a trophoblast protein became recognized as the antiluteolytic substance implicated in maternal recognition of pregnancy in the ewe. A somewhat similar protein, bovine trophoblast protein-1 (bTP-1), also consisting of multiple isoforms, was soon after identified as a product of cow conceptuses (9). As with oTP-1, it is produced during the critical period when the corpus luteum must be rescued if progesterone production by the ovary is to be maintained and the pregnancy continued. Intrauterine infusion of bTP-1 through the cervix of cows between days 15.5 and 21 significantly extended the length of interestrous interval (10) and reduced the pulsatile output of the luteolysin, $PGF_2\alpha$, from the uterus (11).

INTERFERON-LIKE SEQUENCES OF oTP-1 AND bTP-1

The λ-gt_{11} cDNA libraries were prepared from mRNA isolated from ovine and bovine conceptuses and screened with an antiserum to oTP-1 (12–13). Several "full-length" cDNA were identified and subjected to nucleotide sequencing. It soon became clear that multiple kinds of mRNA for oTP-1 and bTP-1 were represented in the libraries and that these mRNA likely arose from different genes. There was, in addition, at least 85% sequence identity between the cDNA for oTP-1 and bTP-1 (13). The cDNA represented mRNA that were about 1 kb in length and possessed a

585-base open reading frame. They coded for 195-amino acid polypeptides with 23-residue signal sequences.

Comparison of the base sequences of the coding regions demonstrated that the cDNA for oTP-1 and bTP-1 shared about 65% identity to a series of cloned cDNA representing interferons of the alpha (or leukocyte) class. The similarity to IFNα was also noted in several laboratories by NH_2-terminal sequencing of the purified proteins (14–16). The latter studies also confirmed the presence of different protein isoforms of oTP-1. The finding that oTP-1 and bTP-1 were most probably IFN-like molecules was startling since IFN, though broadly based in their activities (17–19), had never been implicated directly in events associated with early pregnancy and certainly not as molecules involved in conceptus-maternal signaling.

However, the IFNα that have been most intensively studied are 166 or 165 residues in length and are inducible by virus in leukocytes. They are the products of a large family of intronless genes and are found in all mammalian species so far tested. In the human there are at least 15–20 members of this family clustered together on chromosome 9 (19).

Relatively recently, a second group of IFNα genes was defined by screening genomic (20) or leukocyte cDNA libraries (21) under relatively nonstringent conditions with IFNα cDNA probes. The studies have revealed a family of genes that hybridized only weakly to the probes and coded for IFN that were 6 amino acids longer than the well-studied IFNα. They were called either IFNα-II by Capon, et al. (20) to distinguish them from the better known IFNα-I, or IFNΩ by Hauptmann and Swetly (21), who judged them to have diverged sufficiently from the IFNα to constitute a new family. This issue of nomenclature remains to be resolved, but the term IFNα-II will be used for the rest of this chapter when referring to the 172-amino acid IFN.

The trophoblast proteins oTP-1 and bTP-1 clearly resembled the IFNα-II more than the more familiar IFNα-I both in terms of the lengths of their polypeptides and in amino acid sequence (13). The similarity to the predicted sequence of a bovine IFNα-II was 65%–70% compared to a 47%–54% identity for bovine, human, rodent, and porcine IFNα-I. However, oTP-1 and bTP-1 clearly resembled each other (~80% identity) more than they did the bovine IFNα-II. Nevertheless, they possessed several regions of sequence conserved in all IFNα so far characterized, including the four cysteines that participate in intrachain disulfide bonds and a highly conserved stretch of peptide near the carboxyl terminus. Moreover, the sequence similarity of oTP-1 and bTP-1 to IFNα occurred throughout the lengths of their polypeptide chains and was not confined to specific regions. It should also be recognized that the IFNα-I are themselves a highly divergent group of proteins (22–23). In general, they rarely show more than 60% sequence identity even between species that are closely related in the evolutionary tree. This diversity in sequence and the multiplicity of the genes themselves have raised questions as to whether the different proteins might have subtly different functions. It also remains unclear whether the actions of oTP-1 and bTP-1 on various reproductive parameters can be completely mimicked by other IFNα.

oTP-1 AND bTP-1 AS FUNCTIONAL IFN

The IFN were first described by Isaacs and Lindenmann in 1957 (24) as protein factors that protected cultured cells from viral lysis. Subsequently, three major types of IFN have been identified: the IFNα (or leukocyte IFN); the IFNβ (or fibroblast IFN), which are related to the IFNα in sequence and which bind to the same type 1 receptors; and the unrelated IFNγ (or immune IFN) whose receptors and actions on target cells are very different from the α- and β-types. Both oTP-1 and bTP-1 have potent antiviral activity that approaches that of other known IFNα (25). The specific antiviral activity of a recombinant bTP-1 made in this laboratory (>10^8 IU/mg) is comparable to that of other IFN prepared by such methodologies (19), and oTP-1 is effective against a range of viruses and will act on human as well as bovine cells (16).

All available evidence is consistent with the view that oTP-1 and bTP-1 act through "classical" IFNα receptors. Their binding to crude membrane preparations from ovine endometrium is of high affinity and completely displaceable by recombinant human (14) or bovine IFNα-I (26). Interestingly, however, crosslinking studies have indicated that oTP-1 becomes associated with two polypeptides (Mr 100,000 and 70,000), whereas the bovine IFNα-I only becomes crosslinked to the larger component. The significance of this observation remains unclear, particularly since unlabeled recombinant bovine IFNα-I_1 inhibits the crosslinking of the oTP-1 to both membrane polypeptides (26). It is also of interest that the tissue with the highest concentration of oTP-1/IFNα receptors in the sheep is the endometrium and not, as expected, the spleen. The ovary is also rich in such receptors (27).

ACTIONS ON THE IMMUNE SYSTEM

Rather than simply representing antiviral agents, the IFN have multifunctional properties with pleiotropic effects on their targets (17–18). They also probably act in concert with other cytokines and growth factors to modulate cell activities. The possible role of the embryonic IFN in regulating cells of the immune system is of interest since it is unknown how the embryonic allograft, which may express foreign histocompatibility antigens, survives in the uterus.

Although there is no evidence that oTP-1 and bTP-1 are involved in modulating immune responsiveness of the mother (or even whether the trophoblast expresses histocompatibility antigens and is recognized as foreign at the time oTP-1 and bTP-1 are maximally expressed), the two proteins share the property of other IFNα of being able to inhibit the proliferation of activated lymphocytes (16, 28). These activities are manifested at protein concentrations below 10^{-9}M, which is consistent with mediation through type 1 IFNα receptors. oTP-1 inhibits PHA-induced lymphocyte blastogenesis without reducing the production of the T-cell mitogen, interleukin-2 (IL-2) (28). Thus, the trophoblast IFN may have a local immunomodulatory role by selectively inhibiting the proliferative responses of certain maternal immune cells to IL-2.

However, IFNα have a variety of actions on the immune system, including an ability to stimulate or inhibit proliferation of cells in the monocyte-macrophage lineage, to increase expression of major class I histocompatibility antigens and to activate natural killer cells (18). In view of the recent interest in the possible trophic effects of T-lymphocyte products on conceptus development, these aspects of IFN activity on the immune system should not be disregarded.

CAN THE ACTION OF oTP-1 AND bTP-1 ON REPRODUCTIVE PARAMETERS BE MIMICKED BY OTHER IFNα?

Until recently, when bTP-1 was synthesized by recombinant procedures in this laboratory, no IFNα-II or trophoblast IFN from any species was available in the quantities required for large-scale testing in animals. Even now the bTP-1 we produce is not of pharmaceutical quality. However, a recombinant bovine IFNα-I (rboIFNα-I) has been made available by CIBA-GEIGY Basel, Ltd. In our laboratory 125 μg of such a preparation was introduced into the uteri of ewes twice daily between days 12 and 15 of the estrous cycle. Comparable doses of oTP-1 are able to extend luteal function (8). However, no extension of the cycle was noted with the rboIFNα-I_1 treatment (H. Francis, J.C. Cross, and R.M. Roberts, unpublished results). A comparable study was performed by Stewart, et al. except they used 2 mg daily (29). In this case, cycle length and corpus luteum function were significantly extended in the treated ewes. Such data have suggested that the rboIFNα-I_1 may not be as effective as the natural trophoblast product, oTP-1.

Short extension of the estrous cycle has also been noted in cows when high doses (2 mg) of rboIFNα-I_1 were introduced into the uterus. Of particular interest, however, has been the ability of intramuscular injections of even larger amounts of rboIFNα-I to extend interestrous interval (30). This observation was unexpected because oTP-1 and bTP-1 have been assumed to act locally on the endometrium (5, 7). However, we have recently shown that such injected IFN can influence protein synthesis in the endometrium of sheep (H. Francis, T. Schalue-Francis, and R.M. Roberts, unpublished results) and presumably, therefore, alter other aspects of uterine physiology. Nevertheless, it is possible that the injected IFN has pharmacological effects at sites other than the uterus. An action on the ovary, for example, cannot be excluded.

Even more intriguing have been the effects of intramuscular injection on reproductive performance in ewes. In two well-controlled studies (31–32) in which bred ewes were injected with rboIFNα-I_1 or placebo during the time of maternal recognition of pregnancy, IFN significantly increased the number of ewes diagnosed as pregnant. These results, which have clear commercial importance, need to be examined in more extensive field trials, particularly as there is, as yet, no evidence for comparable effects in cattle. In addition, the trophoblast IFN themselves should be tested for their ability to improve reproductive performance.

INTERFERON PRODUCTION BY CONCEPTUS TISSUE OF OTHER SPECIES

Antiviral activity, indicative of IFN, has been found associated with conceptus tissues of several mammalian species in addition to cattle and sheep. Clearly, this subject is of great interest since it might indicate that IFN has a general role in early pregnancy of all species. In this section, we shall briefly review what is known about the association of IFN with pregnancy in the pig, horse, mouse, and human. We shall also present results from our own laboratory that cast doubt upon a universal role for IFN in maternal recognition of pregnancy.

Pig

In swine a putative IFN is released by peri-implantation conceptuses as early as day 11 of pregnancy and has a molecular weight of about 22,000 (33). The antiviral activity present is several orders of magnitude less than that produced by sheep and cow conceptuses at comparable stages of their pregnancies. In part, this low activity might result from the instability of the porcine IFN. However, the activity is not, as originally thought, associated with the major secretory proteins of the conceptuses, which appear instead to be retinol-binding proteins (34). In addition, we have been unable to detect any mRNA that hybridized with a range of IFN probes that have included ones representing porcine IFNα-I, oTP-1, and mouse IFNβ (J.C. Cross and R.M. Roberts, unpublished results). Attempts to screen a cDNA library from day 11–15 porcine conceptuses with a porcine IFNα-I probe under nonstringent conditions also failed, and we have been unable to neutralize the antiviral activity itself with antisera raised against porcine IFNα-I, oTP-1, human IFNα/β, or murine IFNα/β.

Experiments were also designed to determine whether recombinant bovine IFNα-I_1, which has antiviral activity on pig cells, had any antiluteolytic action in pigs. Maternal recognition of pregnancy occurs in swine between days 12 and 15 of the estrous cycle (5) at the time the embryos elongate from spherical to filamentous forms and begin to secrete measurable antiviral activity (33). In 1982 Ball and Day (35) tested the ability of conceptus homogenates to maintain a unilateral pregnancy when they were infused into the surgically isolated unoccupied uterine horn (from which embryos had been flushed prior to day 8) during the critical period in which corpora lutea begin to regress in the nonpregnant animal. Normally, pigs cannot maintain such a unilateral pregnancy, presumably because the endometrium of the nongravid uterine horn releases sufficient $PGF_2α$ to cause luteal regression on both ovaries. We employed the same model, but infused recombinant bovine IFNα-I_1 (0.33 mg every 8 h) into 5 unilaterally pregnant gilts between days 10 and 15 after previous estrus (day 0). This treatment failed to maintain the pregnancies, and in all gilts the serum progesterone levels began to decline sharply by day 13. Controls ($n = 5$) infused with saline gave similar results. Mean estrous cycle lengths in the two groups were 19.7 ± 1.5 and 19.6 ± 1.2 days, respectively. These experiments suggested that bovine IFNα-I_1 was not antiluteolytic in swine and that IFN might not have a role in maternal recognition of pregnancy.

Horse

Maternal recognition of pregnancy and prolongation of luteal function in the mare, as in other large farm animals, occurs before the conceptus becomes firmly attached to the uterine wall (36). $PGF_2α$ derived from the endometrium, is also considered to be the natural luteolysin in this species (37). By day 14 of diestrus in the cycling mare, peripheral plasma progesterone has already begun to decline, histological changes in luteal cells become evident, and the $PGF_2\alpha$ concentration in the uterine vein has started to increase (37). In addition, luteal function can be extended in mares when blastocysts are removed at day 15 or later, while there is no effect at day 14 or earlier (38). By analogy with cattle and sheep, therefore, it seemed likely that if IFN were antiluteolytic in the mare, they would be produced by conceptuses at around day 14 to 15 of pregnancy. An experiment was therefore designed to test this hypothesis.

Two grade mares were teased to determine the onset of estrus and then bred by natural service at alternate days throughout estrus. In addition, the mares were palpated daily through estrus to detect the day of ovulation (day 0). Fifteen days later embryos were flushed nonsurgically from the uterus with modified Dulbecco's phosphate-balanced saline solution plus 2% (by volume) fetal calf serum. Embryos were transported to the laboratory, washed several times in culture medium, and cultured for 24 h at 37°C under the conditions described by Godkin, et al. (6). Samples of media were removed at 0, 3, 6, 12, and 24 h. These were held at 4°C for 6–24 h and then assayed for antiviral activity (16, 33).

A total of 3 embryos were recovered on day 15 postovulation. Based on palpation, one of the mares appeared to have ovulated 2 follicles 4 days apart. When the uterus of this mare was flushed, 2 embryos were recovered with diameters of approximately 10 mm and 20–25 mm. One embryo was determined to be ~11 days of age (10-mm diameter), while the larger embryo was judged to be approximately 15 days of age. A single 20-mm spherical embryo was recovered 15 days after palpable ovulation from the second mare. Antiviral assay of all culture samples from these 3 embryos yielded negative results in three separate assays and suggested that IFN were not products of conceptuses at this critical stage of pregnancy.

Mouse

There have been numerous accounts of antiviral activity associated with mouse embryo and placenta, although size fractionation of the proteins responsible has indicated that the sizes of the active proteins were unusually large for typical IFN and often incompletely neutralized by antisera to IFNα/β (39). We have demonstrated that preimplantation mouse blastocysts release small quantities of antiviral activity and that such activities persist until at least day 16 of gestation (40). This activity also appeared to be neutralized by antiserum against murine IFNα/β. However, it was not possible to demonstrate the presence of IFNα or IFNβ transcripts in tissues of mouse embryos around the implantation period by using a range of sensitive procedures (nuclease protection assays, cDNA library screening, in situ hybridization, Northern blotting).

Human

Antiviral activity neutralizable by antiserum to human IFNα (but not IFNβ) has been reported in the amniotic fluid from normal human pregnancies (41). IFN-like activity is present in fetal blood, fetal organs, placenta, and decidua and in medium from perfused human-term placenta (42). Several molecular weight and antigenic types of IFN have been identified (43–44), including a 43-kD species neutralized by both IFNα and IFNβ antisera and six additional species (15–80 kD) fully neutralized by IFNα antiserum (43). Immunocytochemical analysis has localized constitutive IFNα, IFNβ, and IFNγ to cells of the syncytiotrophoblast (45–46). It should be stressed, however, that amounts of antiviral activity associated with the human conceptus materials have been very low compared to those noted in ruminants.

In our laboratory we failed to detect IFNα transcripts by in situ hybridization in sections of first-trimester conceptus tissues removed surgically from ectopic sites in the fallopian tube (C.E. Farin and S. Heyner, unpublished results). Nor were we able to detect antiviral activity in culture medium from human embryos cultured in vitro to the blastocyst stage prior to their transfer to patients undergoing IVF procedures (J.C. Cross, unpublished results).

In a more recent study, human cytotrophoblast cells were isolated from first-trimester placentas and studied for IFN production. Briefly, washed fetal villi were treated with collagenase/hyaluronidase to remove the syncytial layer, and the cytotrophoblast cells dissociated with trypsin (47). The cells were purified on Percoll gradients and plated on fibronectin-coated polycarbonate filters (10^6 cells/15-mm-diameter filter). After 18 h of culture in Dulbecco's modified eagle's H21 minimal essential medium containing 2% nutridoma, the medium was aspirated from the cells and assayed for IFN. No significant antiviral activity was detected.

CONCLUSIONS

Together, the results detailed in the previous sections suggest that the production of IFNα by the trophoblast noted in sheep and cattle may not be typical of all other mammalian species. Conceivably, IFN has no role in maternal recognition of pregnancy in nonruminants or is required in much smaller amounts to exert its effects. Indeed, it is puzzling why the sheep conceptus produces so much oTP-1 (48) when type 1 IFN receptors have such high affinities (Kd $< 10^{-10}$M) for their ligands and could presumably operate effectively at concentration orders of magnitude less than those found locally in the sheep uterus. One possibility is that oTP-1 and bTP-1 do not act locally, but must diffuse or be carried to some more distant tissue.

REFERENCES

1. Morton H, Tinneberg HR, Folfe B, Wolf M, Mettler L. Rosette inhibition test: A multicentre investigation of early pregnancy factor in humans. J Reprod Immunol 1982;4:251-61.
2. O'Neill C. Platelet-activating factor and maternal recognition of pregnancy. J Reprod Fertil 1989;37(suppl):19-27.
3. Moor RM. Effects of embryo on corpus luteum function. J Anim Sci 1968;27 (suppl 1):97-118.
4. Betteridge KJ, Eaglesome ND, Randall GCB, Mitchell D, Lugden EA. Maternal progesterone levels as evidence of luteotrophic or antiluteolytic effects on embryos transferred to heifers 12-17 days after estrus. Theriogenology 1978;9:86-93.

5. Bazer FW, Vallet JL, Roberts RM, Sharp DC, Thatcher WW. Role of conceptus secretory products in establishment of pregnancy. J Reprod Fertil 1986;76:841-50.
6. Godkin JD, Bazer FW, Moffatt J, Sessions F, Roberts RM. Purification and properties of a major, low molecular weight protein released by the trophoblast of sheep blastocysts at day 13-21. J Reprod Fertil 1982;65:141-50.
7. Godkin JD, Bazer FW, Roberts RM. Ovine trophoblast protein 1, an early secreted blastocyst protein, binds specifically to uterine endometrium and affects protein synthesis. Endocrinology 1984;114:120-30.
8. Godkin JD, Bazer FW, Thatcher WW, Roberts RM. Proteins released by cultured day 15-16 conceptuses prolong luteal maintenance when introduced into the uterine lumen of cyclic ewes. J Reprod Fertil 1984;71:57-64.
9. Bartol FF, Roberts RM, Bazer FW, Lewis GS, Godkin JD, Thatcher WW. Characterization of proteins produced in vitro by peri-attachment bovine conceptuses. Biol Reprod 1985;32:681-94.
10. Thatcher WW, Hansen PJ, Gross TS, Helmer SD, Plante C, Bazer FW. Antiluteolytic effects of bovine trophoblast protein-1. J Reprod Fertil 1989;37(suppl):91-9.
11. Knickerbocker JJ, Thatcher WW, Bazer FW, Barron DH, Roberts RM. Inhibition of uterine prostaglandin $F_2\alpha$ production by bovine conceptus secretory proteins. Prostaglandins 1986;31:777-93.
12. Imakawa K, Anthony RV, Kazemi M, Marotti KR, Polites HG, Roberts RM. Interferon-like sequence of ovine trophoblast protein secreted by embryonic trophectoderm. Nature (Lond) 1987;330:377-9.
13. Imakawa K, Hansen TR, Malathy P-V, et al. Molecular cloning and characterization of complementary deoxyribonucleic acids corresponding to bovine trophoblast protein-1: A comparison with ovine trophoblast protein-1 and bovine interferon-α_{II}. Mol Endocrinol 1989;3:127-39.
14. Stewart HJ, McCann SHE, Barker PJ, Lee KE, Lamming GE, Flint APF. Interferon sequence homology and receptor binding activity of ovine trophoblast antiluteolytic protein. J Endocrinol 1987;115:R13-5.
15. Charpigny G, Reinaud P, Huet J-C, et al. High homology between a trophoblast protein (trophoblastin) isolated from ovine embryo and α-interferons. FEBS Lett 1988;228:12-6.
16. Roberts RM, Imakawa K, Niwano Y, et al. Interferon production by the preimplantation sheep embryo. J Interferon Res 1989;9:175-87.
17. Rossi GB. Interferons and cell differentiation. In: Gresser I, ed. Interferon 6. London:Academic Press, 1985:31-68.
18. Tamm I, Lin SL, Pffeffer LM, Sehgal PB. Interferons α and β as cellular regulatory molecules. In: Gresser I, ed. Interferon 9. London:Academic Press, 1987;13-73.
19. Pestka S, Langer JA, Zoon KC, Samuel CE. Interferons and their actions. Ann Rev Biochem 1987;56:727-77.
20. Capon DJ, Shepard HM, Goeddel DV. Two distinct families of human and bovine interferon-α genes are coordinately expressed and encode functional polypeptides. Mol Cell Biol 1985;5:768-79.
21. Hauptmann R, Swetly P. A novel class of human type 1 interferons. Nucl Acids Res 1985;13:4739-49.
22. Gillespie D, Pequignot E, Carter WE. Evolution of interferon genes. In: Carne PE, Carter WA, eds. Handbook of experimental pharmacology; vol 71. New York: Springer-Verlag, 1984:45-63.

23. Zoon KC, Wetzel R. Comparative structures of mammalian interferons. In: Carne PE, Carter WA, eds. Handbook of experimental pharmacology; vol 71. New York: Springer-Verlag, 1984:79-100.
24. Isaacs A, Lindenman J. Virus interference, I. The interferon. Proc R Soc Lond (Biol) 1957;147:258-67.
25. Roberts RM. Minireview: Conceptus interferons and maternal recognition of pregnancy. Biol Reprod 1989;40:449-52.
26. Hansen TR, Kazemi M, Keisler DH, Malathy PV, Imakawa K, Roberts RM. Complex binding of the embryonic interferon, ovine trophoblast protein-1, to endometrial receptors. J Interferon Res 1989;9:215-25.
27. Knickerbocker JJ, Niswender GD. Characterization of endometrial receptors for ovine trophoblast protein-1 during the estrous cycle and early pregnancy in sheep. Biol Reprod 1989;40:361-9.
28. Niwano Y, Hansen TR, Kazemi M, et al. Suppression of T-lymphocyte blastogenesis by ovine trophoblast protein-1 and human interferon-α may be independent of interleukin-2 production. Am J Reprod Immunol 1989;20:21-6.
29. Stewart HJ, McCann SHE, Lamming GE, Flint APF. Evidence for a role for interferon in the maternal recognition of pregnancy. J Reprod Fertil 1989;37(suppl):127-38.
30. Plante C, Hansen PJ, Martinod S, Siegenthaler B, Thatcher WW, Leslie MV. Effect of intrauterine and intramuscular administration of recombinant bovine interferon α1 on luteal lifespan in cattle. J Dairy Sci 1989;72:1859-65.
31. Nephew KP, McLure KE, Day ML, Xie S, Roberts RM, Pope WF. Enhancement of maternal recognition of pregnancy and embryo survival in sheep by treatment with recombinant bovine interferon-α_I1. J Anim Sci 1990 (in press).
32. Roberts RM, Schalue-Francis T, Francis H, Keisler D. Maternal recognition of pregnancy and embryonic loss. Theriogenology 1990;33:175-83.
33. Cross JC, Roberts RM. Porcine conceptuses secrete an interferon during the preattachment period of early pregnancy. Biol Reprod 1989;40:1109-18.
34. Harney JP, Mirando MA. Retinol-binding protein (RBP): A major secretory component of the pig conceptus [Abstract]. Biol Reprod 1989;40(suppl 1):131.
35. Ball GD, Day BN. Bilateral luteal maintenance in unilaterally pregnant pigs with infusion of embryonic extracts. J Anim Sci 1982;54:142-9.
36. Sharp DC, McDowell KC. Critical events surrounding the maternal recognition of pregnancy in mares. Equine Vet J 1985;3(suppl):19-22.
37. Sharp DC, Zavy MT, Vernon MW, et al. The role of prostaglandins in the maternal recognition of pregnancy in mares. Anim Reprod Sci 1984;7:269-82.
38. Hershman L, Douglas RH. The critical period for the maternal recognition of pregnancy in pony mares. J Reprod Fertil 1979;27(suppl):395-401.
39. Roberts RM, Farin CE, Cross JC. Trophoblast proteins and maternal recognition of pregnancy. In: Milligan S, ed. Oxford reviews of reproductive biology. Oxford: Oxford University Press, 1990 (in press).
40. Cross JC, Farin CE, Sharif SF, Roberts RM. Characterization of the antiviral activity constitutively produced by murine conceptuses: Absence of placental mRNAs for interferon alpha and beta. Mol Reprod Dev 1990 (in press).
41. Lebon P, Girard S, Thepot F, Chany C. The presence of alpha-interferon in human amniotic fluid. J Gen Virology 1982;59:393-6.
42. Chard T. Interferon in pregnancy. J Dev Physiol 1989;11:271-6.
43. Duc-Goiran P, Robert-Galliot B, Lopez J, Chany C. Unusual apparently constitutive

interferons and antagonists in human placental blood. Proc Natl Acad Sci USA 1985;82:5010-4.

44. Santos JR, Reis LFL, Ferreira PCP, Gomes JAS, Golgher RR. Human amniotic membrane interferon (IFN-mA) preparations contain a new antigenic type of human IFN. In: Abst 1988 meet Int Soc Interferon Res. Kyoto, Japan, 1989:44.
45. Howatson AG, Farquharson M, Meager A, McNicol AM, Foulis AK. Localization of alpha-interferon in the human feto-placental unit. J Endocrinol 1988;119:531-4.
46. Paulesu L, Bocci V. Which is the role of IFN during pregnancy? In: Kawade Y, Kobayashi S, eds. The biology of the interferon system. Proc 5th annu meet of ISIR, Kyoto, Japan, 1988. Tokyo:Kodansha Scientific, 1989:323-7.
47. Fisher SJ, Cui TI, Zhang L, et al. Adhesive and invasive properties of human placental cytotrophoblast cells in vitro. J Cell Biol 1989;109:891-902.
48. Ashworth CJ, Bazer FW. Changes in ovine conceptus and endometrial function following asynchronous embryo transfer or administration of progesterone. Biol Reprod 1989;40:425-34.

Author Index

Subject Index

www.ingramcontent.com/pod-product-compliance
Lightning Source LLC
LaVergne TN
LVHW052305200926
848481LV00002B/470
9781461278191